FROM FALLING BODIES

TO RADIO WAVES

Classical Physicists and Their Discoveries

FROM FALLING BODIES
TO RADIO WAVES

Classical Physicists and Their Discoveries

Emilio Segrè
UNIVERSITY OF CALIFORNIA, BERKELEY

W. H. FREEMAN AND COMPANY
NEW YORK

Cover illustrations taken from text for jacket and paperback editions as follows: Front cover, left to right, top: Galileo (p. 2); Huygens (p. 37), Newton (p. 6); middle: Fresnel (p. 91), Faraday (p. 153), Maxwell (p. 12); bottom: Hertz (p. 175), Carnot (p. 195), Boltzmann (p. 237). Back cover, left to right, top: pp. 62, 257, 56; middle: pp. 92, 110, 206; bottom: 161, 107, 234.

Library of Congress Cataloging in Publication Data

Segrè, Emilio.
  From falling bodies to radio waves.

  Bibliography: p.
  Includes indexes.
  1. Physics — History.   2. Physicists — Biography.
I. Title.
QC7.S435   1984   530'.09   83-16584
ISBN 0-7167-1481-7
ISBN 0-7167-1482-5 (pbk.)

Printed in the United States of America

3 4 5 6 7 8 9 0 MP 4 3 2 1 0 8 9 8 7 6

Contents

instrument invented by mankind." Electromagnetism: the current and the needle — Oersted and Ampère. A fresh look and a powerful imagination: Faraday — from bookbinder to prince of science and experimenter supreme. Infallible in physics: Maxwell. A bridge to modern physics: H. A. Lorentz.

Preface

When I wrote *From X-rays to Quarks*, I wanted to communicate to friends and to a younger generation the impressions gathered during a lifetime devoted to physics. The period treated was one in which physics flourished, and the story had a general cultural interest. I found that more people than I initially thought were interested in the subject, and the lectures I gave to lay audiences and my classes at Berkeley on the history of contemporary physics were unusually popular.

The writing of the book whetted my curiosity about earlier physicists. I have not met any of those mentioned in this volume, but I tried to get acquainted with them through both their scientific and personal writings.

Galileo holds a special position in that company of physicists and for me personally. When I was about fifteen years old, I knew a certain amount of elementary physics, having spent many childhood hours with a text by Adolphe Ganot. I do not know how I got the idea that I could learn more by reading Galileo. I found the *Dialoghi dei massimi sistemi* difficult, and I did not learn much physics from it. On the other hand, the book made a deep impression on me. Galileo was a truly intelligent man who showed how to use one's head. The physics in the book was modest, and it was obtained with great effort, but the intellectual example was impressive. For that reason, I bought a picture of Galileo, which I hung in my room. It is still there. I also read more of his works and his letters, including those he wrote to his daughter. I had lived in Florence for some time, and at the end of my readings and experience, I felt as if I had actually known the man. Of course, this is an illusion, but in any case I formed definite ideas about Galileo, whether right or wrong.

The late General George Patton, according to the movie version of his life, believed in the transmigration of the soul and, while standing amidst the ruins of Carthage, imagined himself to be the reincarnation of Hannibal. I do not believe any of this, but, figuratively, there can be a certain communication with the spirits of the past.

As I read many of the original papers fundamental to physics, I could perceive the difficulties faced and overcome by their authors. Through their works, *we* know how they looked at their problems; what seemed, and is, important; what should be neglected; and, ultimately, what the answers are. *They* did not know and had to find out. That is the great difference between studying a text and studying the "book of Nature." This book is a testimonial to my affection for my scientific forefathers. It derives from the desire to know "Chi fur li maggior tui?" as Dante said, or to quote the title of a contemporary book, to know ones "Roots." It does not pretend to be a history of physics.

Having spent many years as a research physicist and having observed at close quarters quite a few of the great physicists of this century, I know firsthand some of their glories and tribulations, which helps to understand their predecessors. Admiration for the great minds of our scientific ancestors comes naturally, and reviewing their work one marvels at the progress of events and the almost inevitability of their discoveries. From a technical point of view, the sum of their work is a harmonious construction in which each part fits into the next.

But leaving aside the specific discoveries and their applications, the advances are less admirable. The philosophical problems that plagued the Greeks, Descartes, Galileo, Newton, and Leibnitz are still with us. One may try to shrug them off by saying that they seem irrelevant to "scientific" progress. In other words, we may have as complicated an object as a color television set, which allows us to see and hear instantly what happens on another continent, but we still may have doubts about the foundations of science.

That is the public position of the great majority of scientific workers. They may also have their private thoughts, but for various reasons they seem inclined to keep them private. The reasons may range from fear of being burned at the stake to Gauss's fear of the "Larm der Boeoten" ("the noise of the idiots"). It may not be fashionable or prudent to show epistemological worries, but some of the most illustrious physicists have done so, even in recent times.

In that respect, the founding fathers of physics were more candid. Galileo, Huygens, Descartes, and Newton did not hide their preoccupations, including religious ones. When we come to Faraday or Maxwell, we have the impression that they erected an impervious wall between their sincere and deep religious beliefs and their physics.

Most modern physicists are satisfied that their art, as practiced, gives answers that can be validated by experiment and are of great predictive power, technically prolific, and aesthetically satisfying. They leave the deeper roots to the philosophers.

The successes of the philosophers are dubious. Even in such things as the analysis of space and time, I venture to say that mathematicians like the founders of non-Euclidean geometry and physicists like Einstein have achieved more than the professional philosophers. Some constructions, for instance the standard formulation of nonrelativistic quantum mechanics, have left serious doubts in the minds of the great practitioners of physics. The doubts are of a philosophical nature and, beyond physics, metaphysical. It seems to a simple-minded practitioner that philosophy, although often extolled by great physicists, did not really influence them very much. They kept it in a closet outside their laboratories or studies, invoking it ex post facto. The limits of what one can learn, the meaning of "truth," the relation of the mind to the objective world, the definition of the words in this sentence, are all points one could discuss indefinitely.

A palliative to these problems was, smugly, given by Goethe with an accommodating motto: "Das schoenste Glueck des denkenden Menschen ist das Erforschliche erforscht zu haben und das Unerforschliche ruhig zu verehren" (Goethe, Sprüche in Prosa, *Ueber die Naturwissenschaft V*). As translated by W. B. Roennfeldt, this means, "The finest achievement for men of thought is to have fathomed the fathomable, and quietly to revere the unfathomable."

In writing this book I have enjoyed many of my readings. I have tried to include enough of the original writings to give an impression of their authors and of the style of their work and their times. I have also developed certain feelings toward the great men I have described. With many I became a "friend," although in several cases not without considerable awe.

Finally, the reader should know that in venturing so far from my bailiwick, I heard repeatedly the Greek painter Apelles' voice warning: "Ne supra crepidam sutor indicaret," or "A cobbler should not judge above his last." I am aware of the risk of not heeding that exhortation, and I hope the reader will not resent my rashness.

I owe a debt of gratitude to Professors C. W. F. Everitt, R. Hahn, and J. Heilbron for many instructive conversations. Their professional prowess has been illuminating to me on many occasions.

Many institutions and individuals have helped me with the illustrations, and I want to thank them all here, in addition to the credit given at the proper place.

September 1983   Emilio Segrè

FROM FALLING BODIES
TO RADIO WAVES

Classical Physicists and Their Discoveries

A Whimsical Prelude

In my work as a physicist, I've often thought of what it would be like to have met with the great men whose work is described in this book, to have talked to them, seen their laboratories, and moved within the times in which they lived. I have tried to put some of my impressions into a fantastic dreamlike prelude; indeed the following pages report only dreams, some of them real, some imagined.

I traveled to Padua for the first time in the spring of 1610 to meet Galileo. His friend Sagredo had kindly written a letter of introduction. The woman servant who greeted me at Galileo's door bade me enter and told me that Galileo was not in, but was due to return at any moment from a trip to Venice, some twenty-five miles away. As I waited, I noted that the house showed a certain luxury and excellent taste, with elegant Tuscan furniture about and paintings by Cigoli and other artists on the walls. As I moved about, I saw a small cottage near the house. It appeared to be a workshop or laboratory of some sort, and there was a man working within. I beckoned to him and walked over. He introduced himself as Marcantonio Mazzoleni, a skilled artisan, and explained that he was finishing some proportional compasses for sale to Galileo's students. Now, however, the great new enterprise was lens grinding and he devoted most of his time to it. Galileo, too, was spending long hours in the shop grinding and testing lenses. Mazzoleni was guarded as he spoke; he clearly did not want to reveal any trade secrets.

    I left the cottage to return to the house, passing through a beautiful orchard filled with blooming fruit trees and carefully tended vegetable beds. I noted that there was no trace of Galileo's children, nor of Marina Gamba, who, I had heard, lived with him *more uxorio* in a house nearby. I was still looking around when Galileo arrived. He had just returned from Venice's Arsenal, where he had visited with the best craftsmen and had seen some of the operations that especially interested him. He had also paid a visit to the Murano glassmakers and lunched with some important friends. I presented Sagredo's letter, he read it, and when he heard me speak, he addressed me with the familiar "thou." "You sound like a Florentine to me," he

Galileo Galilei (1564–1642), as painted by Sustermans. (Galleria degli Uffizi, Florence.)

said. I explained that although I was born near Rome, in Tivoli, my mother was from Florence, where I had spent many months. Upon hearing that, Galileo grew even more cordial.

He spoke a little about Florence and then passed on to the interesting things he had seen at the Arsenal, commenting on the strength of materials and explaining his ideas concerning ropes and tackles and their uses. I listened and did not dare to interrupt him. Suddenly, he asked, "Do you know how long a dangling iron wire has to be before it breaks under the stress of its own weight?"

"It will depend on the thickness," I answered too quickly. I perceived immediately that I had uttered a stupidity, but Galileo did not give me time to correct myself.

"Blockhead!" he exclaimed. "Don't you see that the thickness of the wire does not affect the answer? However, the case of a heavy wire fixed by two nails in the same horizontal line is much more interesting." He then went into some arguments that I found difficult to follow. The result was that the wire was supposed to take a parabolic configuration. The reasoning was quite complicated, and I was lost before he had finished. Galileo spoke quickly and with great vivacity. He moved incessantly from one idea to another; it was difficult to keep up with him.

In the meantime it had become dark. Very kindly, Galileo invited me for dinner. There were several other students at the table, and we ate very good Tuscan food. He also produced a bottle of wine, a gift of friends who had vineyards in the Colli Euganei, not far from Padua. The dinner, however, did not last long. I could see that he had other things in mind and did not want to waste time at the table. "Come and see the sky with the new glasses," he said. "I finished grinding the lenses last week, and they are really excellent. The glass should magnify surfaces about two hundred times." I would not have dared to ask him for the privilege of looking through his telescope, but I jumped at the invitation. I had never looked through a "perspicillum," as a telescope was called at the time, and I had great curiosity, but I was disappointed, especially at the beginning of the observation. Galileo was full of enthusiasm. The sky had been cloudy for a few days but was now clear; first he looked at the Medicean Stars as he had called Jupiter's satellites. "Look how they have changed," he exclaimed, comparing what he saw with the drawing of his last observation. "They are now on the opposite side of Jupiter." He gave me his place at the telescope. I looked eagerly, but with the best will I could not see all the wonders that Galileo was pointing out to me. The telescope wobbled, and the images were faint and fuzzy. Galileo became impatient with me but benevolently concluded, "At least you are trying your best. With a little practice you will end by seeing. You are not like some asses who do not want to see."

Thirty years elapsed before I revisited Galileo. I climbed to his villa in Arcetri in 1640, and I was saddened by the great change in his appearance. I was confronted with an old man who spoke mainly of the past, not without bitterness, but with great circumspection. He still showed the ancient intellectual fire, but it was dimmed by the insults he had suffered from his enemies and from the Church. He remembered my Padua visit. "Those were good times," he said. But when he turned to scientific arguments, he seemed to recover his old vigor. Two young fellows, Vincenzo Viviani and Evangelista Torricelli, attended him with obvious reverence and love. They tried to substitute for his failing sight, and they hung on his every word. Galileo spoke of his book, *Nuove Scienze,* and of what he planned to add to it in a new chapter that he had only partially completed. "These young men," he said, "offer such excellent criticism that I have to answer them. They will be the worthy followers in this work."

Late in the evening I said good-bye. Descending the steep street of the Costa San Giorgio under a clear full moon, with all the gardens filling the air with the sweet smells of the spring flowers, I thought of my present visit and of the earlier visit in Padua. I knew I had met one of the greatest human minds. I was speculating on what the future might bring us in Galileo's line of thought, but it did not occur to me that one day we would go to the Moon and that his *Nuove Scienze* would evolve into modern physics.

My dreams took me next to Holland, nearly fifty years after my visit to Galileo.

In 1684, I went to The Hague to visit Huygens. Although I do not know Dutch, I had no difficulty in finding his house. Even the mispronounced name was enough to elicit directions. The Huygens residence was a great palace surrounded by a large garden in one of the main squares of the city. Huygens lived there, in the house of his ninety-year-old father, who was known to all as a great writer, diplomat, and public servant. The splendid residence revealed at once the exalted social rank of the inhabitants. I was led upstairs and found a luxurious hall with wooden panels, shining brass at the doors, everything demonstrating diligent household help. After a few minutes Huygens entered the room and addressed me in perfect French, far better than mine. He also said a few words in Italian and asked me about Florence and the Grand Duke of Tuscany. It was obvious, however, that the language to be spoken was French. Seeing me look at the paintings hung on the walls, Huygens asked, "Do you like them? They are truly beautiful, even if they differ from your Italian paintings. Look: this is by my father's friend Rembrandt. Father bought it from him when Rembrandt was very young. I still remember the times when Rembrandt used to visit us at home; I was a child and I listened to the conversations between him and my father. He was a great painter, but in his old age he went out of fashion. What should one say? Even the Dutch who claim to understand painting are subject to fashions and do not know when they are faced with a masterpiece."

We soon turned to physics. I told him of my visits to Galileo at Padua and Arcetri. "Great man," he exclaimed, "the teacher of us all. He too was a friend of my father, who visited him at Arcetri. Galileo changed the art of measuring time." He moved on and showed me one of his clocks and started to explain the mysteries of the cycloidal pendulum. "It is quite elegant as mathematics, but not so practical. I have better hopes for a clock with no pendulum at all, where the isochronous period is given by a spring. In telescopes, too, we have made great progress. Have you seen what was hidden behind the 'Saturno tricorporeo' (he used the Italian words)? It has a ring, and Galileo saw it under different angles but did not succeed in figuring out the true nature of the object. I too have some difficulties in understanding the whole affair, but I am sure that the main feature is a ring. Now, however, I work mainly on optics."

He paused for a moment as if hesitating to broach the subject. "I have found some extremely interesting things, and they are surely true, but the great Newton is making trouble. He is a great man, but quite a character. A few years ago, when he sent me his *Principia,* I was dumbfounded. I had to work hard for many weeks to master the book. However, gravitation is still a mystery. He describes it but does not explain its nature. Do you know Newton? Try to visit him." At this point he looked at me intensely and smiled faintly. I noticed his expression but did not understand why he smiled in that way. "The mathematics Newton uses is also all new. I still follow old Archimedes, although I recognize that between Newton and Leibnitz they have found new methods that the younger generation will have to learn if it wants to make progress."

Drawing by Huygens, 1658. (Courtesy of the Doe Library, University of California, Berkeley.)

Later, his conversation turned again to art and Italy. "My most original countrymen do not want to see Italian art anymore. Can you imagine that Rembrandt refused to travel to Italy as everyone used to do? Our light here is different from that of your country; you do not appreciate our skies, with their clouds and fogs and subdued colors. I will show you some of my small drawings. They're not much, but drawing is one of my hobbies, as is lens grinding. But at lens grinding, I am a master; in drawing, I am not."

We had spent an hour together, and Huygens seemed to have grown quite tired and appeared to me to be in pain. I then remembered the frailty of his health, which had forced him to leave the France he loved and return to Holland. I did not want to abuse his courtesy, and it was clear that he had other things to do. I took my leave, with many thanks and a small drawing he gave me as a souvenir.

In my dream meetings, I could not have missed the great Newton. That dream, however, turned into a nightmare.

I went to Cambridge in 1690 with the hope of meeting Newton. Some friends had cautioned me that the enterprise would not be as easy as I believed it would. Nor could I forget the mysterious smile I had seen on Huygens's lips. Since my friends

Isaac Newton (1642–1727). Portrait by Sir Godfrey Kneller, 1702. (Courtesy of the National Portrait Gallery, London.)

were familiar with Newton's idiosyncrasies, they gave me detailed suggestions on how I should behave, what subjects to avoid, and what people I should not mention. Somewhat alarmed by all these instructions, I decided to observe Newton from afar, before trying to visit him.

I went to Trinity College and waited near the gate of his quarters, half hidden by some bushes. It was not long before he came out. Perhaps because of all that I had heard about him, I was intimidated and uneasy at seeing him. Suddenly,

Sir Isaac turned around and looked straight at me, as if he had sensed I was watching him. Sternly, he addressed me, "Sir, why do you intrude on my privacy? Get out from there at once and be gone." And so I was.

Thus, it was impossible to converse with him, and I had to find a móre mundane way of getting acquainted with Newton. In the real world of 1982, I went to Cambridge and revisited the city and Trinity College, both of which I knew from a sojourn in 1934. Not much had changed, except for the cars and traffic. I went to the university's library, where I found some of Newton's papers classified in neat cartons, mostly from the Portsmouth Estate. Obviously, Newton had preserved many notes and scraps of writing that others would have thrown away. Touching those papers gave me an eerie feeling. To have in my hands a sheet of the manuscript of the *Principia,* written and touched by Newton, was moving. The paper used ranged from a bond with a beautiful watermark, worthy of the Fabriano mills in Italy, to ordinary paper that had survived only because in Newton's time nothing but rags were used for papermaking.

Some of the sheets held writing in a clear, elegant hand, in Latin or in English, according to subject. As I read some of the papers, I could recognize passages from the *Principia,* or versions thereof. Some sheets were so full of erasures and rewriting that they were almost illegible. The beginnings of mechanics had clearly given Newton all the trouble I had expected. Erasure upon erasure, piled on the formulation of the laws of motion, made them very difficult to read.

Another sheet contained some routine mint business in English, followed immediately, in a smaller hand, by mathematics and by a sharp attack on Leibnitz in Latin. Laborious numerical calculations followed, with hundreds of figures filling all the space on both sides of page after page. They seemed to me to be on astronomy. Some of the sheets were half burned, probably from the fire that devastated Newton's study around 1675.

Newton's variable moods were clearly reflected in the papers — the indefatigable persistence in the pages of neat numbers, the doubts and struggles for the exact formulation of the principles of mechanics, the demons of jealousy against this or that presumed offender, the mysterious impulses toward mysticism and alchemical experiments — all were there to see.

My next journey was to an altogether different world. One could see immediately the change in dress, from the ornate eighteenth-century fashions to the bourgeois costume so similar to present-day dress. Of course, fashion was the least of the great changes made by the French Revolution.

I visited Paris in 1810, and after I was settled in my lodgings, attended a meeting of the First Section of the Institut de France, the scientific branch of the Académie des Sciences that was the official meeting place of French science. I found considerable confusion there, with members speaking aloud among themselves, while the official lecturer desperately tried to explain his own work from the lectern to the inattentive crowd. Amidst the confusion it was difficult to approach the famous, but a distinguished looking gentleman smiled at me. I returned his smile, and he came over to talk to me. Although he had a slight accent, his French was much better than mine; I explained to him why I was there and my wish to meet some of the distinguished company. He kindly answered, "Come next Thursday to the villa of the Marquis Laplace at Auteuil."

I was a little taken aback, since how could I, a complete unknown, intrude on the great Laplace, author of *Le système du monde,* chancellor of the Imperial Senate, great cordon of the Legion of Honor. I asked my new acquaintance his name, which I had not understood at first. To my surprise, he was none other than Alexander von Humboldt, a Prussian by birth, but very much at home in France. If he was inviting me, I could assume that I would be well received. Thus, the following Thursday, properly dressed, I left my lodgings near the Place de la Concorde at the center of town. After a pleasant two-hour walk in the picturesque Bièvre Valley, I arrived in the village of Auteuil. When I asked for Laplace's house, I was immediately shown a large building with a garden. I entered the house and was greeted by Humboldt, who then introduced me to the master of the house and his other guests, who seemed to be regular visitors.

The oldest guests were Berthollet, the famous chemist who had worked with Lavoisier about twenty years earlier, and Laplace. They appeared to be in their sixties and were the senior and most respected members of the company. Berthollet and Laplace were distinguished from a group of younger men in their thirties, which included the mathematician Monge, the physicochemist Gay-Lussac, and the youngest of them all, Arago, who was in his twenties. All the guests seemed to be close friends, and they spoke animatedly on subjects familiar to them; their conversation was filled with puns, double meanings, and allusions that I could not grasp. There was a good deal of academic gossip, and the younger members of the group, especially, joked without much reverence about the emperor, "who had not found the time" to read the volumes of the *Mécanique céleste* that the master of the house had humbly dedicated to him. But Napoleon had made Laplace wealthy and covered him with honors, as he had also done for Berthollet, to whom the emperor, according to hearsay, had recently given a large sum of money.

The guests soon came to order. Gay-Lussac gave an interesting report on gaseous combinations — on the volumes of the components and of the products. A lively discussion followed in which the audience criticized the results and proposed corrections to the measurements, delving into minute experimental details. Gay-Lussac answered the criticisms and then moved on to show some of the apparatus and experiments in Berthollet's laboratory. Berthollet lived in the immediate neighborhood, in a villa as luxurious as that of Laplace, in which he had converted several of the rooms into laboratories. Berthollet had balances, barometers, hygrometers, and many other instruments built by his Parisian friend Fortin, who had built the precision balances that Lavoisier used for reforming chemistry. Guests from all countries were received by Berthollet in his private laboratory, and he often let them use his facilities, offering suggestions and advice.

The examination of the apparatus and experiments and the scientific debate, which often grew quite heated, lasted until dinner time. There were many corrections

Joseph Louis Gay-Lussac (1778– 1850), French chemist and physicist, whose studies on the expansion and thermal behavior of gases, on chlorine and iodine, and on the industrial manufacture of sulfuric acid have permanent value. He was a prominent member of French society. (Courtesy of the Doe Library, University of California, Berkeley.)

to the volumetric measurements of gases, for pressure, temperature, possible presence of impurities, and many other factors. Friends and adversaries of Gay-Lussac did not relent. At eight o'clock, we moved to the dining room, where we were joined by the ladies, including Mesdames Laplace, Berthollet, and Arago. The conversation turned from scientific subjects to society news, including the election of new academicians, the appointment of new professors in the major Parisian schools, and the ladies-in-waiting of the empress. The guests held various official positions, and it was clear that their opinions carried much weight in the selection of such persons.

I sat silent as appropriate for a foreign visitor. Mme. Berthollet had discerned my Italian accent in the few words I said when we were introduced, and during dinner very kindly reminded me that in the fall of 1801, Volta had sat at the same table. She then told me of his memorable visit, at which her husband, Laplace, and other scientists would not leave him in peace, so fascinated were they by his new electric pile. Even the emperor, then first consul, had attended all of Volta's lectures at the Institut. She then added, "When Volta came here for dinner, he had to stay the night because they could not stop talking electricity. On the other hand, when His Majesty invited him for dinner, the invitation did not arrive in time, and Volta's place remained empty."

While we chatted, talk among the scientists at the table veered to physicochemistry. What kept the molecules together? Were the molecular bonds related to capillarity? Did they have a range? The Marquis de Laplace tended to pontificate on the subject, but the others did not hesitate to contradict him, even if they did so very respectfully.

Michael Faraday (1791–1867), after a painting by Thomas Phillips.

I journeyed next in time to London of 1846, for I could not miss the opportunity to visit Faraday.

Tickets for the Friday evening lectures at London's Royal Institution on Albemarle Street cost a guinea each, an expense that kept me on a light diet for several days. I had hoped the speaker would be Faraday, but to my disappointment the scheduled lecturer was Wheatstone. The sum I had spent for the ticket was considerable, and it would have taken me some time before I could afford another. Through a strange chain of events, however, everything turned out for the best.

The lecture was to have started at eight o'clock and I knew that time was kept very precisely, but no one appeared at the podium for several minutes after the appointed time. Then, to my great surprise and satisfaction, Faraday entered and took the podium. I recognized him immediately, even though I had never met him. He began to talk as if nothing was amiss, simply saying a few words of apology for the announced speaker, and then turned to the announced topic — Wheatstone's electric chronoscope. I was impressed by his ability to improvise a lecture. Following his own "Advice to a lecturer" he spoke "slowly and deliberately, conveying ideas with ease from the lecturer and infusing them with clearness and readiness into the minds of the audience." I perceived a trace of cockney accent in his pronunciation. He also performed some experiments with complete success and obvious pleasure for himself and the public.

However, he finished his talk about twenty minutes before the end of the allotted hour; then, appearing somewhat reluctant, he changed his tone and said that he would now confide some tentative thoughts on the nature of light, which he had always kept to himself. Faraday then became less clear in his exposition, and I doubt that he told all that he had in his mind. First, he spoke in general terms of the unity of natural forces, but he soon descended to particulars and affirmed that he thought the speed of light in a medium had something to do with its dielectric constant.

The next day, I summoned all my courage and went to visit Faraday at the Royal Institution. Although it was a Saturday, he was at work on an experiment, and I felt uncomfortable about wasting his time, but he set me at ease with the courtesy of a true gentleman. Seeing him at close quarters, I was even more impressed than during the lecture by his charm, physical beauty, graceful movements, and, above all, the extraordinary penetration of his eyes.

Faraday showed me around the Institution, including the basement room where frogs were kept, possibly a reminder of Galvani's time and experiments. What seemed to interest Faraday most were some pasteboards on which iron filings traced the lines of force of a magnet. He showed me in detail how to obtain such images and how to fix them so that they would persist after the magnet was removed. "The key to magnetism is in the understanding of such natural figures," he said. "They say much more than the formulas. Here there is everything, but I cannot yet understand them completely. I think constantly of them; sometimes I even dream of them."

I tried to turn the conversation to the lecture of the previous evening, but I immediately regretted it. Although he retained his perfect politeness, Faraday clearly indicated that he did not want to talk on that subject. I also felt that it was time to leave. The experiment on which he was working interested him much more than my visit, and I did not want to abuse his courtesy. Leaving the Royal Institution in a thick fog, I was almost run over by a cab, so deep was I in thoughts of my visit to Faraday, the unique person I had had the good fortune to meet.

Remembering my lack of success in trying to see Newton during my first Cambridge visit, I was a little reluctant to return there. But, in 1876, I had heard so much about the new Cavendish Laboratory and its influence on British education, as well as about the Cavendish Professor, J. C. Maxwell, that I decided to journey back to Cambridge.

I met Professor Maxwell at his house, and we walked together to the laboratory, followed by a dog named Toby, who seemed to have an uncanny way of communicating with his master. Maxwell, after some formal civilities, sank into deep thought, totally oblivious to my presence. Several people greeted him as we walked to the lab, but I had the impression that he did not see them, so deeply was he absorbed. The same absentmindedness was displayed toward the laboratory's janitor.

Trinity College was little changed, particularly the little bush that had served as my hiding place while I observed Newton. But the Cavendish Laboratory in Free School Lane was brand new, even though it was built in the gothic style. I followed Maxwell into his study and, after a few minutes, he came down from the clouds and asked me in a very friendly way what I wanted to see. I noticed that his pronunciation

James Clerk Maxwell (1831–1879). (Courtesy of C. W. F Everitt.)

was different from the usual Cambridge accent. He rolled his "r" almost as much as I did, which pleased me, but he also spoke very fast, and I had the impression that he did not care whether or not I could follow him. Occasionally, he would jump from one subject to another without any apparent connection. He also used unfamiliar words and made puns, at which he laughed maliciously, and I stupidly, because I could not understand their meaning.

Maxwell took me next to the students' laboratory. It was not very well equipped, with the majority of the instruments of the "string and sealing wax" type. The realization of absolute electrical standards of units was one of the main research subjects for the students, and the professor himself worked on the same theme but with much more refined apparatus. Maxwell gave me a short lecture on the methods he was using. Almost all of them had ingenious compensation features, and several were null methods. He had a special knack for avoiding first-order errors; although the resulting calculations were far from simple, the professor seemed to enjoy them.

He suddenly changed the subject and started to talk about kinetic theory. Here things became more complicated; he went to a blackboard, which he rapidly covered with equations, trying to explain to me several questions relative to specific heats of polyatomic molecules and soon moved to gaseous diffusion. Suddenly, he stopped, and smiled skeptically. "Here there is real trouble," he said, "and it must be of a malignant nature. I have racked my brain on it for a long time, but I cannot make progress. Kinetic theory is beautiful and *true,* but it also contains some real puzzles. Why does it not give the right specific heats? They derive so simply from hypotheses that seem indisputable and that, moreover, work so well in many other cases. I have a feeling that until this mystery is solved, one will not understand the true nature of molecular dynamics.''

I understood only part of what Maxwell said, and I even suspected that he was poking fun at me. His conversation, unusual because of his Scottish accent, the breadth of the subjects he broached, and the depth of his comments revealed a great mind, but the visit had left me confused. I was sure that most of his thoughts had been hidden from me. As I left the laboratory, I saw Trinity College once again, and I thought how peculiar these Cambridge eccentrics are who create so much new science.

My final journey in time to visit one of the great physicists would take me to Berlin, a journey I did not relish, especially since I had heard that the city in 1890 resembled a barracks in which the new Kaiser ruled with little tact and even less good taste. But that had little to do with His Excellency Geheimrat Professor Doktor Hermann von Helmholtz.

German science had proved particularly hospitable to young scientists of other countries, and visitors were allowed to take advantage of the well-organized teaching and of the facilities of the rich German laboratories. This being Prussia of 1890, I donned a black suit with striped pants, a stiff collar, and gold cuff links with a small sapphire in the center of each, which I had borrowed from my father. The costume made me slightly uncomfortable, like a uniform on someone not accustomed to wearing one, but as soon as I entered the Reichsanstalt I knew I had been right to wear such attire.

My appointment with His Excellency was for nine o'clock in the morning and I was taken to his study precisely on the hour by a uniformed employee.

Hermann von Helmholtz (1821– 1894), as photographed around 1890, when he was president of the Reichsanstalt in Berlin and the most authoritative German physicist of the time. (Courtesy of Leonard B. Loeb, L. Loeb Collection, Physics Department, University of California, Berkeley.)

Helmholtz was finishing a scientific conversation with an assistant; a quick look around suggested to me that as a younger man I was well advised to stand, almost at attention, until I was invited to sit down. Helmholtz rapidly finished his instructions, the assistant clicked his heels and vanished.

His Excellency then turned to me most affably and asked me if I preferred to speak German or English. I chose German as a sign of deference to him, and he immediately inquired with obviously sincere interest about my work. I perceived that the Prussian crust was melting and was being replaced by a genuine benevolent interest and a desire to help. As I spoke, Helmholtz would interrupt with detailed penetrating questions and would warn me of some subtle cause of error that I had to guard against. He also suggested the means to do so. I had the impression that he had meditated at length on the subject (some experiments on electromagnetic waves), and that he knew a great deal about them.

Helmholtz next inquired about some of his Italian friends, in particular about Pietro Blaserna and commented on his way of playing the harmonium. That brought him to a musical and acoustical diversion, so much so that he went to a harmonium that he had in his study and demonstrated Tartini's sounds to me. By then it was a quarter to ten, and the uniformed employee discreetly opened the door and stuck his head in the room; I got the signal that my time was up and that I better leave quickly. His Excellency invited (or did he order?) me to return the next day for the physics colloquium. He promised to introduce me to other young physicists and would look into how I might work in the laboratory. He also promised to suggest some subject of common interest for my work.

As I left the Reichsanstalt, I encountered a battalion of goosestepping soldiers on parade in one of the great Berlin *Allées*. I had a difficult time trying to reconcile the sight before me with the scientific part of my interview with Helmholtz. It was a world with two faces looking in opposite directions. I shuddered at the thought of what that might bring about.

Chapter 1

The Founding Fathers: Galileo and Huygens

When did physics begin? That is not an easy question. Technological prowess is very old, and the people who built aqueducts or raised pyramids had to know a certain amount of what we now call physics, although they would not have recognized it as such. They were writing prose without knowing it, like Mr. Jourdain, the Molière character. Applied technology however, is not conscious physics. The Greeks developed a highly sophisticated mathematics, and Archimedes statics could be called physics even by modern standards, but they did not establish a doctrine. To trace early physics is beyond my competence, and I take advantage of the obvious discontinuities presented by the development of physics as a science to start my story with Galileo.

The times immediately preceding Galileo are full of precursors. Astronomers, navigators, artists, and technologists raised practical and theoretical questions that are pregnant of the future. Many of the great artists of the Italian Renaissance — Leonardo da Vinci, for example — had an insatiable curiosity for what we would now call scientific problems. However, I find a great difference between their surmises, occasionally truly far-seeing, and the results of work done about a century later in Galileo's time. Mental attitudes and methods had rapidly changed. We can recognize this if we make any attempt to read the "scientific" literature of the late fifteenth or sixteenth centuries, even though we would not find much that we could call scientific in a modern sense. The capital methodological discoveries were made by Galileo when he recognized the power of the combination of experiments with mathematics. A scholar would qualify the attribution of this discovery to Galileo, but it has sufficient truth in it to justify our starting our study with him.

Pisa: Preparation

Galileo Galilei was born in Pisa of Florentine parents on February 15, 1564. The family was ancient and prominent in Florence, but by the time of the birth of

Galileo, it was not wealthy. His father was a musician of considerable reputation; in fact, his music is still played occasionally today.

Galileo passed the first ten years of his life in Pisa, went to Florence around 1574, and was back in Pisa in 1581, registering as a student of medicine at the university. When he was nineteen years old he became acquainted with geometry by reading books and meeting the mathematician Ostilio Ricci (1540–1603). I can very well imagine what a revelation the discovery of geometry must have been for the young man. He was studying something probably distasteful to him, and all of a sudden he found the intellectual activity for which he was born and which somehow had escaped him previously. Probably only passionate love can equal the strong emotion aroused by such an event. He had to argue strongly with his father to obtain permission to pursue his new studies, for the usual reason that they were not practical. This is a very common beginning in the life of scientists.

Soon thereafter he discovered or rediscovered the isochronism of the pendulum oscillations. He wrote some unimportant notes on physics and astronomy. Note that when I say physics, I am not referring to a science resembling our present-day physics. What Galileo called physics at that time has only the name in common with physics in the modern sense, a physics that had still to be created, largely by Galileo himself, some decades later.

Galileo did not finish his medical studies but returned in 1585 to Florence, where he remained for four years occupying himself with diverse studies, chiefly literary. For instance, he gave lectures on the configuration and location of Dante's *Inferno*. He also wrote some papers on hydrostatics and published some theorems on the center of mass of solids. Those papers earned him a certain reputation among the experts, including the Marchese Guidobaldo del Monte of Urbino (1545–1607), who was then and later an important protector of Galileo. Through recommendations he obtained a three-year appointment at the University of Pisa as professor of mathematics. Thus, he returned to the university he had left earlier, his studies incomplete.

He held that post from 1589 to 1592. In those years he apparently started to study the Copernican system, which had been proposed about fifty years earlier by the famous Canon of Thorn (1473–1543); however, he kept the results of these studies to himself. At the same time he wrote more notes on mechanics. They did not contain very important results, although they already indicated the directions in which the mind of Galileo was to work. Specifically, he indicated that he would use mathematics in the study of natural phenomena: mathematics not in the sense in which some of his contemporaries, such as Johannes Kepler (1571–1630), used it—trying to look for harmonies in creation—but rather as an instrument for the quantitative and consistent discussion of concrete problems.

When his Pisa appointment was not renewed in 1592, he found himself without a job and in serious financial straits because the death of his father had placed on him heavy responsibilities for his mother, sisters, and brothers.

Again through the intervention of Guidobaldo del Monte, he obtained a job at Padua and settled there for eighteen years, starting in 1592. Those were the best years of his life, as he acknowledged in old age in writing to a friend. Padua was the university of the independent and rich republic of Venice. It was an old and famous university, and Galileo, despite a very meager salary (180 florins at the beginning), found there a congenial atmosphere. He soon rented a large house where he sublet rooms to students and in which he adapted a room as a shop, for which he hired a worker. This was probably the first embryo of a scientific laboratory with a technician. At Padua he taught geometry and astronomy (the level of the mathematics was not much beyond our present high-school level, and astronomy was according to Ptolemy). In the merry quiet of Padua, Galileo pursued several most important investigations, among them an investigation on the motions of accelerated bodies. He also meditated on astronomy. Private letters indicated that he became convinced of the correctness of the Copernican system around 1597. He invented an instrument, the proportional compass, which is useful for several graphical constructions; in fact, his shop produced a number of these instruments, which he sold at a good profit.

Although there is evidence that most of his discoveries in mechanics matured in Padua, it is remarkable that in this period of great activity — at the prime of his life — Galileo published very little. However, he became well known among the astronomers and natural philosophers of the time. He also struck up close friendships with several gentlemen, most notably Giovanfrancesco Sagredo (1571–1620), a gentleman of considerable wealth and influence in Venice.

In 1599 Marina Gamba, a Venetian woman, came to live with Galileo. They never married, although she and Galileo had three children — a son, Vincenzo, and two daughters. The firstborn, later Suor Maria Celeste, is one of the sweetest feminine figures of all times. Her letters to her father are moving documents of great beauty in their simplicity. Although in 1599 Galileo's salary was raised to 320 florins, he was still plagued by financial obligations deriving from his brothers and sisters. His sister Virginia had married Benedetto Landucci of Florence in 1591. Galileo's father died the same year, and Galileo and his brother Michelangelo were left with the obligation of paying Virginia's dowry. Michelangelo was a musician, rather shiftless, although with some ability, but he did not contribute his share, and the burden fell on Galileo, who for many years was in straitened circumstances. Galileo's mother, Giulia Ammannati, was another problem for the family because she seems to have been of a very difficult character.

I will skip the professional quarrels incurred in Padua, except to remark that Galileo had a very sharp tongue likely to make bitter enemies. We must remember that the scientific style and polemical tone of the seventeenth century were quite different from those employed now; still, the following sample of his style from *Il saggiatore* ("The Assayer") illustrates his manner of attack. Galileo speaks of a pamphlet called "Philosophical and Astronomical Libra" (after the constellation Libra), written by one of his enemies:

> Much more appropriately and truly, if we look at this writing, should he have called it the "Astronomical and Philosophical Scorpion," a constellation called by our supreme poet Dante *"figura del freddo animale"*—"*Che colla coda percuote la gente*"—

. . . "the cold animal which stings people with its tail." And indeed there is no scarcity of stings against me, and even heavier than those of scorpions, inasmuch as these, as friends of man, do not bite unless they are first offended and provoked whereas he bites me who never even thought of annoying him. But lucky for me that I know the remedy and antidote against such stings. I shall squash and rub the scorpion itself on the wounds so that the poison reabsorbed by the corpse of the animal shall leave me free and healthy.

Galileo Galilei, in an engraving executed around 1613, when Galileo was forty-nine years old. Note the two angels at the top of the frame, one with a telescope, the other with a proportional compass. This engraving appears in a booklet Galileo wrote on solar spots. (Courtesy of the Bancroft Library, University of California, Berkeley.)

Despite his stinging wit, Galileo had very loyal and admiring pupils and friends. All this emerges from the very human correspondence that is still extant.

Padua: Marvels of the Skies

The year 1609 was fateful for Galileo's life. Having heard of the invention of the telescope, he reconstructed such an instrument. He called it *perspicillum* in Latin, *occhiale* in Italian — the name "telescope" was later given by Federico Cesi (1585 – 1630) — and turned it first to terrestrial objects. The practical importance of the new invention was immediately apparent to him, and he demonstrated it to several of his influential friends. On August 24, 1609, he wrote to the Doge and Senate of Venice pointing out the military importance of his discovery: "Looking through it, what is nine miles away seems to be at a distance of only one mile: this can be of inestimable value for any business or enterprise at sea or on land. At sea one can discover vessels or sails of the enemy at much larger distance than usual so that for two hours we can see the enemy before he sees us and being able to see the number and quality of his ships we can judge his forces and prepare ourselves to give chase, to fight, or to run away. . . ." The Doge and Senate deliberated and decided that there was no point in trying to keep the invention secret. Furthermore, in sign of satisfaction, they gave Galileo tenure for life at Padua and raised his salary to 1,000 florins per year. That was an unprecedented salary.

When he turned the telescope to the sky, the things he saw were memorable. In rapid succession he discovered Jupiter's satellites, the stellar nature of the Milky Way, the phases of Venus, the strange configuration of Saturn and its "satellites," the solar spots, the mountains of the moon, and other marvels of the sky.

These astronomical discoveries made a tremendous impression on Galileo and on his contemporaries. He says, *"Alcune osservazioni le quali col mezzo di un mio occhiale ho fatte nei corpi celesti; e siccome sono di infinito stupore così infinitamente rendo grazie a Dio che si sia compiaciuto di far me solo primo osservatore di cosa ammiranda e tenuta a tutti i secoli occulta."*—"Some observation of the celestial bodies which I have made with the help of my 'glasses'; and since they are infinitely stupendous I am infinitely thankful to God who has deigned to make me the first observer of things so admirable and hidden to all past ages."

Galileo summarized his astronomical discoveries in a little book, *Sidereus nuncius*. The title was translated by him in Italian as *Avviso astronomico* ("Astronomical Notice"). It shows all the strong feelings of the discoverer. Here are a few sentences from its beginning:

> Great indeed are the things which in this brief treatise I propose for observation and consideration by all students of nature. I say great, because of the excellence of the subject itself, the entirely unexpected and novel character of these things, and finally because of the instrument by means of which they have been revealed to our senses.
> Surely it is a great thing to increase the numerous host of fixed stars previously visible to the unaided vision, adding countless more which have never before been seen, exposing these plainly to the eye in numbers ten times exceeding the old and familiar stars.

Copy of a letter Galileo wrote to Belisario Vinta, secretary to the Grand Duke of Tuscany, announcing his astronomical discoveries in enthusiastic terms. (Biblioteca Nazionale, Florence.)

It is a very beautiful thing, and most gratifying to the sight, to behold the body of the moon, distant from us almost sixty earthly radii, as if it were no farther away than two such measures—. . . .

But what surpasses all wonders by far, and what particularly moves us to seek the attention of all astronomers and philosophers, is the discovery of four wandering stars not known or observed by any man before us. Like Venus and Mercury, which have their own periods about the sun, these have theirs about a certain star that is conspicuous among those already known, which they sometimes precede and sometimes follow, without ever departing from it beyond certain limits. All these facts were discovered and observed by me not many days ago with the aid of a spyglass which I devised, after first being illuminated by divine grace.

The drawings in the *Sidereus nuncius* showing the positions of the satellites of Jupiter have been recently checked against modern tables, and naturally they were found to be correct. More interesting, they have given us a way to estimate the resolving power of Galileo's telescope. It could resolve about twice the diameter of Jupiter (about 4×10^{-4} radians or 1').

Despite his attachment to Padua, Galileo longed to return to Florence for several reasons. He preferred to be relieved from teaching duties, he was probably homesick, and he had family reasons for returning there. The new Grand Duke of Tuscany, Cosimo II de'Medici, offered him excellent conditions, and Galileo accepted. His intimate friend Sagredo warned him about what he might be losing in a beautiful and prophetic letter; in it, one can see the better judgment of the experienced man of affairs. Sagredo extolled the academic freedom of Padua and concluded, "Where will you be able to find the freedom and independence you enjoy in Venice? I am also worried that you will be in a place where the authority of the Jesuits is so important."

Besides his academic position, Galileo left in Padua Marina Gamba, and on his return to Florence, he put his daughters in a nunnery. Even taking into account the customs of the times, this is a dark page in his life.

Florence: Glory and Downfall

The astronomical observations of Galileo convinced him that he had irresistible proofs of the Copernican hypothesis, and he decided to come into the open and to advocate it publicly. The story of his position in this matter is very complicated. To me, the most likely interpretation is that Galileo was really a good Catholic, sincerely attached to his religion. He saw that if the Church supported as an article of faith an astronomical system that was clearly in contradiction to observational facts, it would end by losing the respect of mankind. He therefore made a sincere effort to align the Church on the side of science. By this he meant the new science, which he clearly saw coming, as contrasted to medieval doctrines and methods he felt were doomed. Thus, he tried eagerly and in good faith to convince Rome of the correctness of his observations and of the Copernican doctrine. The difficulties he met may seem strange now, but one can understand why they were very real then. For instance, the

Roman astronomers had to be convinced that the things they saw through the telescope were not optical illusions produced by the instrument. The legitimacy of their doubt is best shown by the fact that Galileo considered it seriously himself, dispelling it by experiment. Furthermore, the whole question of the relation of sensory experience aided by instruments, to "reality" is by no means simple; and Galileo meditated deeply on the problem, reaching conclusions that are still valid.

In 1611 he went to Rome to show his discoveries to the highest Church authorities. He had friends among the cardinals, and even the Pope received him most honorably. He became a member of the newly founded Accademia dei Lincei and participated in their work. On the whole, this visit to Rome was successful in convincing the Church astronomers of the truth of his observational discoveries. These discoveries strengthened immeasurably the Copernican theory, which at that time had not been condemned. There were certain symptoms of uneasiness, but Galileo took a rather optimistic view of the situation and left Rome pleased with the results of his visit. The Tuscan ambassador, writing to the Grand Duke, confirmed Galileo's favorable appraisal.

Back in Florence, Galileo advocated publicly the Copernican doctrine and gradually became the object of virulent attacks from Dominican friars and other members of the Florentine clergy. Galileo explained his position in several public letters to prominent clergymen and to the Archduchess Christina, a member of the Medici family. He attempted a clarification of the position of religion and science on this basis: First, the Scriptures are always right. However, they need interpretation because in their literal expression they contain things that are obviously wrong and even heretical. The manifest errors of the literal texts are due to the historical setting in which they were composed and to the nature of the public to whom they were addressed. Scriptures should not be used in arguments involving experimental facts and their mathematical consequences. When the latter are well established, Scriptures should be interpreted according to our sensory experience and its consequences. "Natural phenomena are directly willed by God. Therefore observations or direct conclusions from them should not be doubted because Scriptures seem to contradict them, because not all the Scriptural sayings are commanded by such strict laws as all natural effects." The truth of the Copernican system does not affect the truth of Scriptures, but only their interpretations—interpretations that are human and do not come from the Holy Ghost.

That position was not liked by the Church, and, finally, in 1616, Galileo was summoned to Rome. He went reluctantly and was coldly received in some quarters, a little better in others. The Florentine ambassador, whose guest he was, anticipated trouble and warned him. In fact, he was brought before the Holy Office and told that the Copernican opinion was condemned and that severe penalties would be imposed if he continued to support it. A moot point is whether the Holy Office told him that he was forbidden to defend Copernicus *quovis modo* (in any way), or not—this became important in his trial of 1633. Galileo was obviously dismayed by this defeat of his hopes; however, he made obeisance. A few days later, on May 26, 1616, he obtained a letter from Cardinal Roberto Bellarmino, stating

Frontispiece of the first edition of Dialogue Concerning the Two Chief World Systems. *Salviati, on the right, sustains the Copernican view; Simplicius, in the center, sustains the Ptolemaic doctrine; and Sagredo, on the left, is the interested listener. (Courtesy of the Bancroft Library, University of California, Berkeley.)*

that he had not been forced to retract anything and absolving him from any accusation of heresy. Soon thereafter he returned to Florence.

Galileo temporarily abandoned his attempt to influence Rome, and devoted all his activity to scientific pursuits, waiting for better times in which to make another attempt to defend his views. In those years he wrote on comets; he also wrote the polemical book *Il saggiatore*, well-known as an outstanding sample of his polemical and literary powers in a subject less technical than his great scientific books.

In 1623 a new Pope was elected, the former Cardinal Maffeo Barberini, who took the name of Urban VIII. He was a close friend of Galileo, to whom he had even addressed a laudatory ode. With new hope, Galileo undertook a major work—the celebrated *Dialoghi dei massimi sistemi* ("Dialogue Concerning the Two Chief

World Systems"). It took him six years to finish the book, which was ready in 1630. The dialogues present three persons: Salviati, who, to a large extent, represents Galileo himself; Sagredo, the open-minded intelligent layman; and Simplicio, an Aristotelian philosopher rather narrow in outlook and occasionally even slightly ridiculous. (Filippo Salviati and Sagredo were two dear friends who had died in their primes, the first from Florence, and the other the Paduan mentioned in connection with Galileo's resignation from Padua.) Besides their main theme, a discussion and comparison of the cosmological systems, the dialogues contain a great number of very clever observations and reflections on various scientific subjects. Today we are used to the methods of science from early schooling, but it is most interesting to see how, for Galileo, almost any daily occurrence presented unexpected problems when

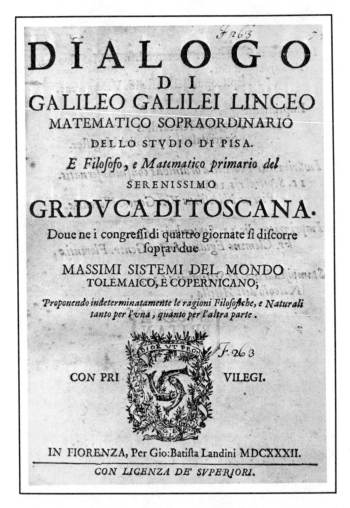

Title page of the Dialogue. (*Courtesy of the Bancroft Library, University of California, Berkeley.*)

seen from the new point of view of the scientist. It is here that the extraordinary importance and greatness of Galileo impress me most. Of course he was ignorant — any schoolboy now knows much more than he did — but his approach was as modern as that of an Einstein or a Rutherford.

A comparison with his great contemporary Kepler brings this out clearly. Kepler, in many ways, was a better astronomer than Galileo: he discovered the elliptic orbits, while Galileo believed them to be circular; he discovered the laws that justly bear his name. But he seems to me incomprehensible in motivation and approach. Although Galileo had Kepler's books, he never read them. This is surprising, but I can imagine that he may have been repelled by Kepler's mysticism. In modern times, Galileo is readable, while Kepler is not. Their different approaches emerge clearly in their use of mathematics. Galileo uses it in a completely modern way to establish, for instance, the motion of a uniformly accelerated point. Kepler wants to connect in a mystical way the radii of the planetary orbits with the regular bodies of solid geometry. (This may be supermodern, because Kepler would certainly enjoy the eight-fold-way and the SU_3 group of modern particle physics.)

Once the book of dialogues was finished, it was essential that it have the Imprimaturs of the Church, which were not easy to obtain. Galileo went once more to Rome. Some changes in the book were requested, which Galileo granted, and a Preface and Conclusion were agreed upon. These are rather startling, at least to a modern reader. In the Preface, Galileo says that the purpose of the book is to show that the Church, which had even consulted him, was well-informed on all Copernican arguments and did not condemn the Copernican system out of ignorance. He states further that he presents the Copernican system only as a hypothesis, fully aware that it is wrong. The agreed Conclusion was also intended to placate the Church. In it Simplicio says, with reference to an explanation of tides which Galileo thought corroborated the Copernican system:

> I admit that your thoughts seem to me more ingenious than many others I have heard. I do not however consider them true and conclusive; indeed, keeping always before my mind's eye a most solid doctrine that I once heard from a most eminent and learned person, and before which one must fall silent, I know that if asked whether God in His infinite power and wisdom could have conferred upon the watery element its observed reciprocating motion using some other means than moving its containing vessels, both of you would reply that He could have, and that He would have known how to do this in many ways which are unthinkable to our minds. From this I forthwith conclude that, this being so, it would be excessive boldness for anyone to limit and restrict the Divine power and wisdom to some particular fancy of his own.

To this argument, which had once been propounded by Urban VIII himself, Salviati and Sagredo bow immediately, declare themselves convinced, and the book finishes. Note that the argument was put in the mouth of Simplicio, who is the Aristotelic philosopher proved to be wrong many times in the dialogues. With these changes the Imprimaturs were granted in 1631.

It is rather strange that anybody could expect to avoid the worst by such simple verbal incantations. In fact, in 1633 the storm broke. The theologians won

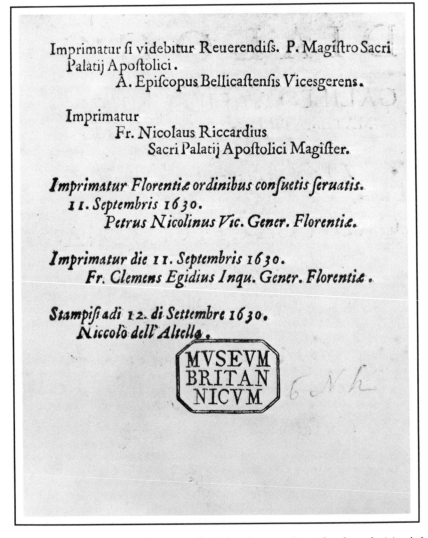

Imprimatur fi videbitur Reuerendifs. P. Magiſtro Sacri
Palatij Apoſtolici.
　　　　A. Epifcopus Bellicaſtenſis Vicesgerens.

Imprimatur
　　Fr. Nicolaus Riccardius
　　　　Sacri Palatij Apoſtolici Magiſter.

Imprimatur Florentiæ ordinibus confuetis feruatis.
11. Septembris 1630.
　　　Petrus Nicolinus Vic. Gener. Florentiæ.

Imprimatur die 11. Septembris 1630.
　　Fr. Clemens Egidius Inqu. Gener. Florentiæ.

Stampiſi adi 12. di Settembre 1630.
Niccolò dell'Altella.

MVSEVM
BRITAN
NICVM

Galileo took precautions by obtaining official imprimaturs from Church authorities before publication of the Dialogue. *(Courtesy of the Bancroft Library, University of California, Berkeley.)*

the Pope to their side, and Galileo, then seventy years old, was summoned from Florence to Rome. He tried to avoid the appearance in Rome, but the Vatican was adamant and threatened to have him arrested. The Grand Duke of Tuscany could not or did not want to protect his mathematician against the Pope, and Galileo went. He was supposed to be jailed, but the Holy Office agreed at first that he could remain in the villa of the Tuscan ambassador Niccolini, a sincere friend of Galileo.

Later, however, the Holy Office insisted on having him in custody. The trial lasted from April to June, 1633, and ended with the complete condemnation of Galileo: he was forced to abjure and to declare on his knees, "Therefore, desiring to remove from the minds of your Eminences, and of all faithful Christians, this strong suspicion reasonably conceived against me, with sincere heart and unfeigned faith I abjure, curse, and detest the aforesaid errors and heresies, and generally every other error and sect whatsoever contrary to the said Holy Church. . . ."

It is a legend that, on hearing the sentence, Galileo stamped the ground and said, *"Eppur si muove"* ("And yet it moves"); but it is a fact that in his own copy of the condemned dialogue he wrote in the flyleaf, "Take note, theologians, that in your desire to make matters of faith out of propositions relating to the fixity of sun and earth you run the risk of eventually having to condemn as heretics those who would declare the earth to stand still and the sun to change position — eventually, I say, at such a time as it might be physically or logically proved that the earth moves and the sun stands still."

If Galileo's condemnation was a tragedy for him, it was also a severe blow for the Church. By condemning what proved to be undoubtedly correct opinions, invoking for this condemnation Scriptural authority and religious arguments, the Church placed itself in an embarrassing position and subsequently lost prestige. This lesson is still remembered, if one may judge by the proceedings of the Vatican Council. Pope John Paul II himself has repeatedly called for a revision of Galileo's trial. Clearly, to maintain a condemnation of his opinions is ridiculous. It has been remarked, however, that the attitude of the Church toward science should not be gauged by the Galileo question, but by more modern issues. In this connection, the Church has assumed an attitude that is extremely similar to Galileo's opinion on the relations between science and religion. In 1979, on the occasion of a commemoration of Einstein, Pope John Paul II said:

> Galileo formulated important norms of an epistemological character, which are indispensable to reconcile Holy Scripture and science. In his letter to the grand-duchess mother of Tuscany, Christine of Lorraine, he reaffirms the truth of the Scriptures: "Holy Scripture can never lie, provided, however, that its real meaning is understood. The latter — I do not think it can be denied — is often hidden and very different from what the mere sense of the words seems to indicate" (*National Edition of the Works of Galileo*, vol. V, p. 315). Galileo introduces the principle of an interpretation of the sacred books which goes beyond the literal meaning but is in conformity with the intention and the type of exposition characteristic of them. It is necessary, as he affirms, that "the wise men who expound it should show its real meaning."
>
> The ecclesiastical magisterium admits the plurality of the rules for the interpretation of Holy Scripture. It teaches expressly in fact, with Pius XII's encyclical *Divino afflante Spiritu,* the presence of different literary styles in the sacred books and therefore the necessity of interpretations in conformity with the character of each of them.
>
> The various agreements that I have mentioned do not in themselves solve all the problems of the Galileo affair, but they contribute to creating a starting point favourable to their honourable solution, a state of mind propitious to the honest and loyal solution of old oppositions.

Galileo was dismayed by his condemnation, but he could do nothing except try to escape to a Protestant country or to a strong republic, such as Venice, which could not be intimidated by the Pope. He decided to abide by the sentence and to return to Florence. When his return was delayed by the plague, he spent several months in Siena as a guest-prisoner of Archbishop Ascanio Piccolomini. Here he was treated with the utmost respect; in fact, the Archbishop was denounced to Rome because apparently "he had suggested that Galileo had been unjustly condemned by the Holy Office which could not nor should not try to confute opinions which had

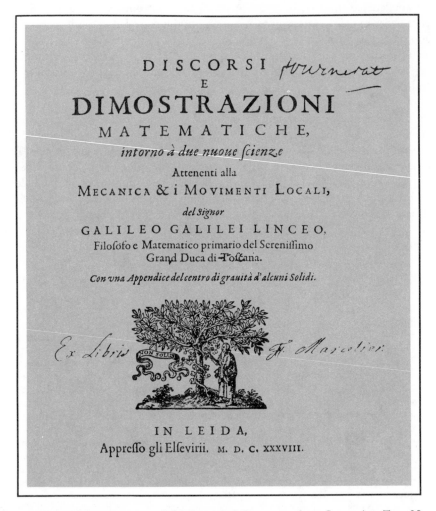

The title page of Discourses and Mathematical Demonstrations Concerning Two New Sciences. *Published in 1638. In Leyden, in order to circumvent the prohibition issued by the Holy Office, it was the last work Galileo published.*

The villa "Il Gioiello" (The Jewel) in Arcetri, near Florence, as it appears today. Galileo was confined in this villa from 1633 on. (Museo di Storia della Scienza, Florence.)

been proved in an invincible way by true and mathematical arguments." This denunciation is in the Vatican Archives.

The period in Siena helped Galileo to recover in part from the blow of his condemnation, and with incredible resilience, at the age of seventy-one, he resumed his scientific investigations and started to write what became his scientific master-piece, *Discorsi e dimostrazioni matematiche intorno a due Nuove Scienze attenenti alla Mecanica e i Movimenti locali* ("Discourses and Mathematical Demonstrations Concerning Two New Sciences"). His correspondence reveals, however, that the contents of that book were already in his mind during the Padua period.

In December 1633 he returned to Arcetri near Florence. There, in April 1634, a most severe blow befell him. Suor Maria Celeste died. Galileo described her as "a woman of exquisite brains, singular goodness, and most affectionate to me." Anyone who has read even a few of her one hundred and twenty-four letters to her father can easily realize what this must have meant to him. The other children were unlike the oldest daughter. The second daughter was a rather obtuse and quarrel-some nun; the son caused much anxiety to his father. Despite everything, the "Two

New Sciences" was finished in 1636. The publication presented great difficulties because it was forbidden in Catholic countries. Galileo passed the manuscript to a French gentleman who had it published by Elzevir in Leyden, allegedly without the permission of the author (who had delivered the manuscript personally to him)!

The "Two New Sciences" is written part in Italian and part in Latin. The persons of the dialogue are those in the *Dialoghi dei massimi sistemi:* Salviati, Sagredo, and Simplicio. The book contains four chapters, or days, to which two more were added later. The first two days deal with problems of strength of materials, including similarity laws and models. This is essentially the first new science. The chapter, however, is rich in digressions on various arguments. Cohesion of solids and water, the nature of a vacuum, pumps, and pendulum oscillations are some of the subjects treated. Extensive reasonings on mathematical problems of what we would now call infinitesimal calculus and set theory are included. There are occasional errors, for instance in the explanation of the limit of the altitude at which a pump can lift water. The limit is ascribed not to atmospheric pressure, but to the cohesion of water. Nevertheless, any modern reader must be impressed by the sharpness of the arguments, by the reference to ingenious experiments, and by the overall intellectual power of Salviati and Sagredo.

The third and fourth days are written in part in a formal way, specifying axioms, theorems and so on, as in Euclid's work. Those parts are in Latin. The subject is uniform motion with constant acceleration and motion of projectiles (including the effect of medium resistance). The third and fourth days contain the law of inertia, the composition of motions according to vector addition, and the study of uniformly accelerated motion. Although the law of inertia is perfectly clear, it is not formally stated as a principle. This has given rise to some dispute as to whether or not Galileo should be credited with it; however, the examples and applications seem to me more important than a formal statement, and it is apparent that Galileo really understood it. After Galileo, nobody questioned the Galilei transformation until the time of Einstein. The third day of the "Two New Sciences" is the highlight of Galileo's work, an achievement of supreme importance to the future of science. It contains a clear understanding of acceleration and the laws of dynamics for a constant force. The fourth day contains, as an application, the parabolic motion of projectiles. The remaining days are more mathematical in character. It is clear that in order to obtain the results mentioned above, Galileo had to have several of the ideas of the calculus. In particular, to understand velocity and acceleration of nonuniform motions, he was confronted with problems of infinitely small quantities, or "indivisibles," as he and his pupil Bonaventura Cavalieri (1598–1647) called them. He succeeded in obtaining the correct answers by geometrical methods.

Shortly after the completion of this book, another catastrophe struck Galileo. He became blind as a consequence of glaucoma. In 1638 he wrote to his friend Diodati, "Galileo your dear friend and servant, since one month is completely and irreparably blind; thus that sky, that world, that Universe which I with my marvelous observation and clear demonstrations have increased in size by hundred or thousand fold, more than believed by all learned men of past centuries, now for me is so limited and restricted that it is not greater than the space occupied by my own body."

But even those misfortunes did not induce the Church to relax its rigorous stand against Galileo. When in 1639, six years after his condemnation, Galileo again applied for freedom, the Pope denied him. However, the young Vincenzo Viviani (1622–1663), an extraordinarily gifted youth of seventeen, was allowed to live at Arcetri as a pupil, secretary, and companion to Galileo. Viviani later wrote a life of Galileo and was one of the first practitioners of the new science of physics. He became first mathematician to the Grand Duke of Tuscany, as Galileo had been before him. In his last years, Galileo had extensive dealings with the Netherlands, concerning a method he had invented to determine longitude at sea using the satellites of Jupiter; however, the method never became practical. In connection with these studies, he invented a form of pendulum clock. In 1639 he extolled the power of the pendulum in determining time and improving astronomical observations. But Galileo's life was approaching its end. He still enjoyed scientific discussions, and he was fortunate to have another young man, Evangelista Torricelli (1608–1647), join Viviani. Torricelli was a star of the first magnitude; it suffices to mention his invention of the barometer. Galileo died on January 8, 1642, in Arcetri.

I have traced briefly the life of Galileo. I have mentioned some of his scientific accomplishments. In my opinion, his discoveries in physics are even more important than his great astronomical findings. It is easy to point out specific discoveries or inventions, but this would give a very incomplete idea of his importance. The method and the spirit of his approach to Nature are his supreme achievements. In the most unexpected places, one finds marvelous things: for example, his attempt to measure the velocity of light (he found that it went faster than he could determine by his method), or his observations on the propagation of sound and on the tones emitted by vibrating strings. And there are many others. The really striking features of his approach to Nature are the open-mindedness of his observations, the sharpness of his reasoning, and his curiosity for all sorts of natural phenomena. Viviani said that Galileo owned few books because he preferred to look at Nature rather than read books. The depth of his thought is shown by his clear understanding of the law of inertia and the invariance of phenomena observed in systems in uniform translation with respect to each other, and by his understanding of the composition and superposition of movements. These are among the most profound scientific insights ever achieved. In his reasoning we often find arguments based on time reversal or other symmetry properties that are strangely reminiscent of some of the most modern considerations of theoretical physics.

Galileo is above all a scientist and not a philosopher. In this he differs sharply from his great contemporary René Descartes (1596–1650), who was preoccupied with metaphysical questions of what to base physics on. Galileo has been considered a symbol of very different philosophical tendencies — platonism, positivism, and so on — because in his writings one can find support of various doctrines, even contradictory ones. In fact, his main point is probably the separation of science from philosophy—hence, his hostility to Aristotle as an authority, not as a scientist. A famous passage in *Il saggiatore* sets forth essential points of his method:

> Philosophy is written in this grand book, the universe, which stands continually open to our gaze. But the book can not be understood unless one first learns to comprehend the language and read the letters in which it is composed. It is written in the language of mathematics, and its characters are triangles, circles and other geometric figures without which it is humanly impossible to understand a single word of it; without these one wanders about in a dark labyrinth.

This is expressed as a counterpart to the use of authority in questions of "natural philosophy." The root of the matter is experience, and mathematics is the proper tool to describe experience quantitatively. Furthermore, mathematical reasoning is to an extent protected from logical errors because the methods of geometry have been well established for a long time.

An evaluation of Galileo is extremely complex. He was undoubtedly one of the most intelligent men who ever existed, if by intelligence we mean analytical ability and scientific imagination. Of his scientific personality, Einstein, a qualified expert, says:

> The *leitmotif* which I recognize in Galileo's work is the passionate fight against any kind of dogma based on authority. Only experience and careful reflection are accepted by him as criteria of truth. Nowadays it is hard for us to grasp how sinister and revolutionary such an attitude appeared at Galileo's time, when merely to doubt the truth of opinions which had no basis but authority was considered a capital crime and punished accordingly. Actually we are by no means so far removed from such a situation even today as many of us would like to flatter ourselves; but in theory, at least, the principle of unbiased thought has won out, and most people are willing to pay lip service to this principle.
>
> It has often been maintained that Galileo became the father of modern science by replacing the speculative, deductive method with the empirical, experimental method. I believe, however, that this interpretation would not stand close scrutiny. There is no empirical method without speculative concepts and systems, and there is no speculative thinking whose concepts do not reveal, on close investigation, the empirical material from which they stem. To put into sharp contrast the empirical and the deductive attitude is misleading, and was entirely foreign to Galileo.

He was very versatile: he is not only one of the greatest physicists that ever existed, but he is also a writer of the first magnitude. He is the greatest prose writer of Italy between Machiavelli and Manzoni, that is, in a span of four hundred years. His musical ability was equal to that of a good professional. Although, in my opinion, he was above all a physicist—the first physicist in the modern sense of the word—the astronomical discoveries of 1610 diverted him to the field of astronomy, and from there he was almost necessarily induced to become a leader in the much broader field of scientific philosophy and of the emancipation and separation of science from religion. His literary gifts made him one of the greatest vulgarizers and popularizers

SAGGI DI NATVRALI ESPERIENZE

FATTE NELL' ACCADEMIA

DEL CIMENTO

SOTTO LA PROTEZIONE

DEL SERENISSIMO PRINCIPE

LEOPOLDO DI TOSCANA

E DESCRITTE DAL SEGRETARIO DI ESSA ACCADEMIA.

SECONDA EDIZIONE.

PROVANDO E RIPROVANDO

IN FIRENZE,

Nella Nuova Stamperia di Gio: Filippo Cecchi. MDCXCI.

CON LICENZA DE SUPERIORI.

The frontispiece of the Saggi di naturali esperienze (Essays on Natural Experiments). *The varied experiments in this now-famous report were performed by the Accademia del Cimento under the protection of the Grand Duke of Tuscany.*

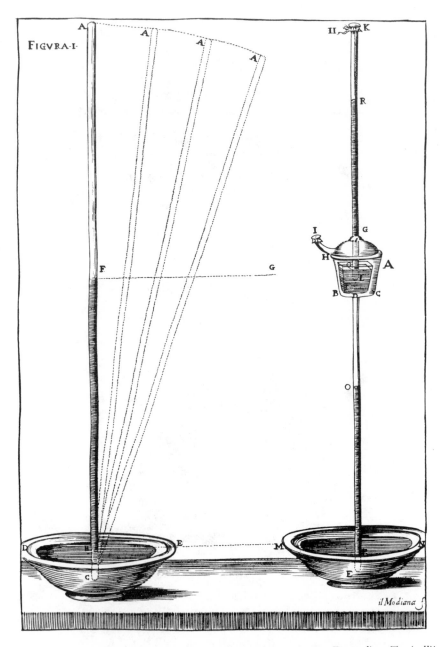

One of the first experiments in Saggi di naturali experienze *concerns Evangelista Torricelli's barometer. This drawing shows an apparatus composed of two barometers, which were used to demonstrate Torricelli's explanation that the height of the mercury column was due to atmospheric pressure. The space above the mercury surface is, by definition, a vacuum.*

of science. His ability as a polemist made him dangerous to the Church and created bitter personal enemies. Yet, his joviality and rather merry disposition attracted many devoted and admiring friends who stood by him even in his disgrace. He was a very hard worker, and although he was unwilling to be a martyr or a hero he had an indomitable spirit.

In Italy, Galileo had a number of direct disciples: Benedetto Castelli (1578–1643), Cavalieri, Viviani, and Torricelli, each of them a notable scientist in his own right. This could have been the beginning of a great school. However, the condemnation of Galileo by the Church had a fatal effect in Italy. His books were all indiscriminately banned, and any scientific activity in Galileo's tradition was at least suspicious. It is true that in 1657 a group of nine men founded the Accademia del Cimento in Florence (the journal *Nuovo Cimento,* known to most physicists, is named after it), which published an interesting book, recording the experiments performed by members of the academy. But it did not survive very long. Ten years after its inception, it was disbanded for various reasons, including the fact that its patron, Prince Leopold of Tuscany, became a cardinal. The termination of the Accademia del Cimento seems to have been a condition of his appointment. The work of the Accademia was important because it showed clearly the power of the new scientific methods. Among the instruments they used were a prototype of a thermometer and one of the first barometers. The members, all gentlemen of leisure, made them with their own hands in many cases, a departure from the custom of the time. This also was probably inspired by Galileo, who repeatedly stressed the importance and interest of technical achievements. For instance, he begins "Two New Sciences" by referring to the great arsenal of Venice (mentioned also by Dante) and to his interest in the skills of the foremen of that institution.

However, the climate in Italy had become unfavorable to science. The leadership passed to the northern nations of Europe, where the state was stronger than the Church. It is almost symbolic that the year Galileo died, Newton was born, and that the Accademia del Cimento was disbanded five years after the foundation of the Royal Society.

Huygens: A Transition Figure

Galileo had struggled in old age with the important problem of the determination of longitude at sea. That problem boiled down to the making of a clock that could keep time on a ship with adequate precision. Galileo's first solution was to use Jupiter's satellites as a clock, but that turned out to be impractical. In connection with those endeavors, he corresponded with a Dutch gentleman, Constantijn Huygens (1596–1684), who elicited great sympathy for the old, semi-imprisoned savant. Constantijn was a rich man, a famous poet, and an important civil servant, serving as an ambassador and a state councillor. He followed the family tradition, because Constantijn's father had also been a prominent man in the affluent and flourishing Dutch society that in many ways inherited and propagated to northern Europe the splendor of the Italian Renaissance.

Constantijn, however, could not suspect that one of his five children, Christiaan (1629–1695), would be the first to practically solve the problem of the clock and, more important, would be the bridge between the times of Galileo and of Newton and one of the great physicists of all times. It is only the circumstance of coming between two such giants and the relative inaccessibility of his voluminous works that make him appear as a transition figure.

Christiaan Huygens was born at The Hague in 1629. Within a circle of a few miles lived Rembrandt (1606–1669), Frans Hals (1580–1666), and Baruch Spinoza (1632–1677), and he or his father had met them all. Having all the advantages of birth, he profited first by being carefully educated by his distinguished father and later by studying at the University of Leyden, where he had Frans van Schooten (1615–1660) as professor of mathematics, himself a son of the previous professor of mathematics at Leyden and an important mathematician whose books would be eagerly studied by Newton.

At Leyden Huygens was strongly influenced by the philosophy of Descartes, his father's friend and frequent guest. However, he recognized after a while that Descartes's physics was something very different from his philosophy; his general rationalistic ideas were valuable, but his particular mechanical views were incorrect. In that period, as did many of his contemporaries, Huygens learned to grind lenses, an art he sometimes practiced with his brother, and he acquired manual dexterity and ability as a mechanic. His telescopes were among the best in existence, and with one of them he discovered a satellite of Saturn and then recognized that the "arms" of Saturn, previously known, were a ring. He wrote on the subject in 1655, but at

René Descartes (1596–1650), the "founder of modern philosophy," in a portrait by Frans Hals. Born in France, Descartes spent most of his life in Holland. His powerful intellect left permanent results in philosophy and mathematics, and his logical methods, if not his results, pervasively influenced physics. (Courtesy of the Louvre, Paris.)

A portrait of Christiaan Huygens (1629–1695), from his collected works. (Courtesy of the Library, University of California, Berkeley.)

first concealed his discovery, as was not uncommon at the time, in an anagram of the sentence "It is surrounded by a thin flat ring nowhere touching and inclined to the ecliptic."

In 1655, at the age of twenty-six, having finished his studies at Leyden, where he had also cultivated legal subjects, he was sent to Paris by his father. His paternal connections in the world of culture helped Huygens, who started to become known for some mathematical tracts and for the Saturn observations. He thus had no difficulty in meeting the important French scientists, and he was informed of the work of Blaise Pascal (1623–1662), Gérard Desargues (1593–1662), and other mathematicians. On his return from France to The Hague, he improved further on his telescopes and his Saturn observations. The period from 1650 to 1666 was the most fertile in Huygens's life. He carried on several investigations at a time, or in rapid succession, on very different subjects. He improved the pendulum construction, inventing a method of obtaining oscillations of a period strictly independent of their amplitude, while in ordinary pendulums the independence is approximate and limited to small amplitudes of oscillation. That investigation is more of theoretical interest than of practical importance, but it produced some elegant and profound mathematics that was to influence much later work. Most important in the period from 1652 to 1656, Huygens investigated the laws of collision, using relativity arguments and establishing for them the conservation of momentum. He also studied centrifugal force (1659), arriving at its value for circular motion. Clocks, pendulums, and the measurement of time were a prime interest for Huygens during

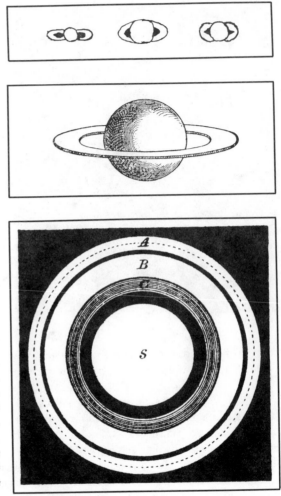

The unraveling of Saturn's rings. Galileo was the first to note something strange about Saturn. The planet appeared to be composed of a central blob, with two small stars touching it on each side. In 1610 he sent Kepler an anagram, as was then fashionable, which when deciphered said, "Altissimum planetam tergeminum observavi" ("I observed that the highest planet is triple"). However, in 1613 he found Saturn to be perfectly round: "What can one say of such metamorphoses? Perhaps the smaller stars vanish like the solar spots? Perhaps Saturn has eaten up his children? Or the sights with lenses have cheated me and so many who have observed with me for so long, and are only illusions and frauds?" In 1616, Galileo drew the top figure in a letter to other astronomers, marveling at the differences between present and previous observations.

By 1655 Huygens, using better telescopes, was able to solve the mystery. He, too, composed an anagram to keep the discovery secret while he continued his studies. Deciphered, the anagram said, "Annulo cingitur tenui, plano, nunquam cohaerente ad eclipticam inclinato" ("It is girdled by a thin flat ring, nowhere touching, inclined to the elliptic"). The middle figure shows one of Huygens's illustrations of the ring as it appeared in Systema Saturnium, which was dedicated to Leopold of Tuscany.

More details about the rings became evident with improved instrumentation, and the question of their nature continued to intrigue astronomers. Maxwell proved that the rings could not be solid but were composed of minor fragments. The bottom figure is taken from his paper of 1856, in which he analyzed the problem.

Saturn's rings as photographed by the unmanned spacecraft Voyager 1 in 1980. (Photo courtesy of NASA and the Jet Propulsion Laboratory.) Voyager was about 8 million kilometers away from Saturn (the distance between Saturn and Earth is about 1,500 million kilometers). The photographs of Saturn's rings, which were made from above and below the plane of the rings, reveal a great deal of complexity. Ring particles seem to have a typical diameter between 1 cm and 1 m, but a few of the small moons have a diameter ranging from 30 to 200 km. These moons are important for determining the size of the gaps between rings. Many other features of Saturn — such as electromagnetic fields, gas and dust distribution, plasma waves, and x-ray emission — could not have been dreamed of even by Maxwell.

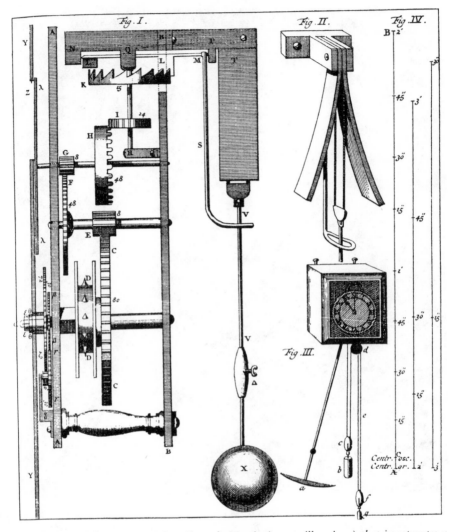

A drawing of a Huygens timepiece (from the Horologium oscillatorium) *that incorporates a cycloidal pendulum that is rigorously isochronous, independent of the amplitude of oscillation, as Huygens had proved. The practical realization of the clock did not match its theoretical refinements, and it was not yet adequate for determining longitude at sea.*

his whole life, especially their practical application to navigation, and in that connection he explained the latitude effect on the pendulum period.

In 1660 Huygens returned to Paris for a second visit. He met Pascal that time and was presented to King Louis XIV. By now he was famous and his British colleagues invited him to London for a visit in 1661. On his return to The Hague, he elaborated on some of the hints received during his foreign trips on various subjects, such as clocks, acoustics, and the nature of a vacuum.

In 1664 he returned to Paris. Jean Baptiste Colbert was finance minister, and he showed great interest in Huygens's work, offering to support it generously. Two years later, on the founding of the Royal Academy of Sciences, he was offered a high pension, a private apartment, and a laboratory. Huygens proceeded to complete much of his previous work in Paris under the auspices of the academy. He established several important theorems of mechanics and in particular, he passed from the mechanics of a point to that of a rigid body. He considered in particular a pendulum formed by a body of arbitrary shape with a fixed axis and was thus led to the concept of what we now call moment of inertia. Huygens helped also a great deal with the organization of the Académie des Sciences.

Huygens participated in the brilliant society life in Paris, but this did not prevent him from accomplishing an enormous quantity of work. His collected

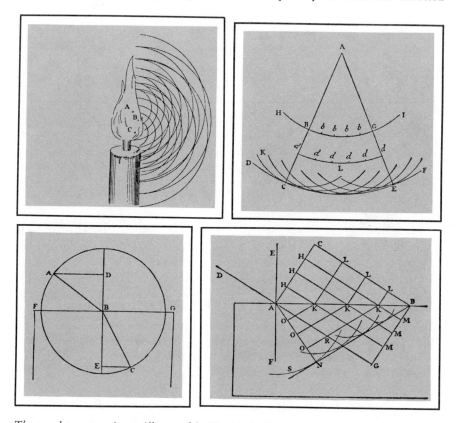

The wavelet construction as illustrated by Huygens in his Traité de la lumière. *The principle that the wave surface could be constructed as an envelope of wavelets emanating from all points of a previous surface gave Huygens a powerful conceptual tool for developing wave optics. It allowed him to explain rectilinear propagation of light, reflection, refraction, and double refraction. Fresnel improved materially on Huygens's principle (1818), but a rigorous mathematical foundation was given only in 1883 by Kirchhoff. (Courtesy of the Bancroft Library, University of California, Berkeley.)*

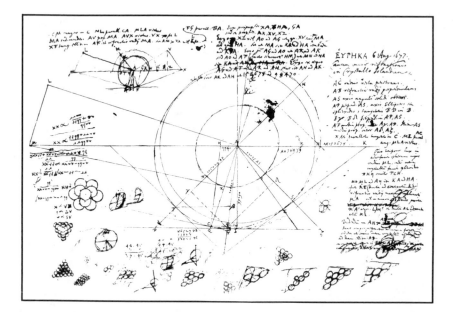

"Eureka!" On August 6, 1677, Huygens found how to explain double refraction, on the assumption that the wave surfaces in Iceland spar are not spheres but ellipsoids (that is, that the velocity of propagation of light depends on direction). The sketches of piles of balls, or ellipsoids, are attempts to find a molecular model for an anisotropic medium. (Courtesy of the Bancroft Library, University of California, Berkeley.)

papers, including his correspondence, fill twenty-two volumes. By 1672 Huygens was experimenting on the double refraction of calcite, which had been discovered in 1669 by the Danish scientist Erasmus Bartholin (1625–1698). Huygens had formulated a wave theory of light—one of his major accomplishments—which explained the laws of reflection and refraction. The fundamental idea was that each point of a wave surface became the center of a new wave and that light manifested itself only on the envelope of all these wavelets. Huygens's theory was sound, but it lacked a clear notion of interference and phase relations. He also thought that the vibrations, similar to sound, were longitudinal. He had reached by analogy and intuition a fundamental result, very difficult indeed to prove, but that, once admitted, gave the key to very recondite phenomena. By using it in 1677, he succeeded in explaining double refraction. In his manuscript dated August 6, 1677, there is a diagram explaining the mystery and a notation in Greek, "Eureka," reminding us of Archimedes jumping out of his bath startled by the discovery of the law of buoyancy. Newton had written important optics papers in 1672, and Huygens had mildly criticized them. That was enough to bring about some typical Newtonian outbursts. The question discussed had nothing to do with the fundamental conception of light, corpuscular or wavelike, and is of little interest now.

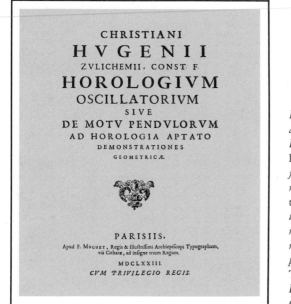

CHRISTIANI
HVGENII
ZVLICHEMII, CONST. F.
HOROLOGIVM
OSCILLATORIVM
SIVE
DE MOTV PENDVLORVM
AD HOROLOGIA APTATO
DEMONSTRATIONES
GEOMETRICÆ.

PARISIIS,
Apud F. Muguet, Regis & Illuſtriſſimi Archiepiſcopi Typographum,
vià Citharæ, ad inſigne trium Regum.
MDCLXXIII.
CVM PRIVILEGIO REGIS.

Huygens was interested in clocks and related problems all his life. In 1658 he published the Horologium. *He returned to the subject with the publication of a major work,* Horologium oscillatorium *(1673), of which this is the title page and which contains many important results in mathematics and physics, including the period of small oscillations* $T = 2\pi\sqrt{l/g}$. *(Courtesy of the Bancroft Library, University of California, Berkeley.)*

Huygens had enjoyed good health in his youth, but from age forty on, he had severe bouts of illness — in 1670 he was ill for several months, from 1676 for almost two years, and again in 1681. Each time he returned to The Hague to cure himself, but otherwise lived in Paris, even when France was at war with Holland. In 1683 Colbert died, and his death brought about a change in French politics, culminating in 1685 with the revocation of the Nantes Edict, which had allowed a certain degree of toleration for Protestants. The king, by a policy of repression, deprived France of many of its best minds, which reemerged in subsequent times as refugees in Protestant countries. That is the reason why we find French names among German mathematicians, poets, and intellectuals. Huygens left Paris and returned to his native country. There he compiled books reporting his previous research. Most famous are the *Horologium oscillatorium* (1673) and his treatises on light, *Traité de la lumière,* published in 1690, and another that appeared after his death. They are in some way the counterpart to Newton's *Opticks.*

During his last illness in 1695, Huygens refused the spiritual assistance of a Protestant minister. That fact was criticized by people of different faiths. Huygens had an anticonformist attitude. He came from a privileged class and had always been able to look down on human frailties. He was diffident, in a time of fierce struggles, of any dogmatic attitude and very aware of human frailty.

Huygens's immense work extended to mathematics. He is one of the precursors of the calculus, and in that field he is also a harbinger of Newton. (I have omitted a number of important contributions that are now incorporated in the standard teachings of mathematics and physics.) He also had ideas on the theory of the origin of gravity that used the Cartesian notion of vortexes. Although incorrect, they are the first hints of the point of view of a propagated action in contrast to an action at a distance.

Huygens's mathematics is quite sophisticated in the use of Archimedean methods. The difference between him and Galileo is very great, and he solved problems, such as those of the tautochrone or of the properties of the evolute of the cycloid, that are way beyond what Galileo could have done. On the other hand, he did not enter into the spirit of the time until very late in years. He did not use notations or ideas originating in that period that revolutionized mathematics.

In any case, Huygens represents one of the high points of the scientific tradition of the Netherlands, a country that has an extraordinary record, considering its continuity, quality, and number of scientists, together with the small size of the land.

Chapter 2

The Magic Mountain: Newton

Carl Friedrich Gauss (1777 – 1855), one of the great mathematicians of all time and a logical mind, chose to have only one "summus," and bestowed that title upon Newton. Poets in many languages have celebrated him, from the verses of Pope,

> Nature and Nature's laws lay hid in night:
> God said, *Let Newton be!* and all was light.

to those of Wordsworth,

> Where the statue stood
>   Of Newton with his prism and silent face,
>   The marble index of a mind for ever
>   Voyaging through strange seas of thought, alone

to those of Halley, Voltaire, Foscolo, and others. They all express awe for such an almost superhuman genius. The same feeling is expressed in his epitaph:

> *Sibi gratulentur mortales tale tantumque*
> *existisse humani generis decus.*

> Let the mortals rejoice that such and so great
> an honor of mankind ever existed.

The following sketch tries to give the reasons for this admiration. I must, however, add that Newton has remained to me, despite my efforts, rather unfathomable.

Newton's life can be divided into three periods: from birth to his arrival at Cambridge in 1661 (nineteen years); his Cambridge period, the thirty-five years from 1661 to 1696, spent entirely at Trinity College; and his London period, the thirty-one years from 1696 to his death in 1727.

A Complex, Mysterious Personality

Isaac Newton was born on December 24, 1642, the year of Galileo's death. He was the son of a yeoman father who had died three months before Isaac's birth. Isaac was born prematurely and was so small at birth that his mother feared for his survival. When he was three years old, his mother remarried and the child was entrusted to his grandmother. Both families lived within two miles of each other, near Woolsthorpe, in Lincolnshire, England. When the boy was eleven, his mother's second husband died, and he returned to live with her, his two half-sisters, and his half-brother.

It is possible that these disturbing shifts in his infancy may have affected Newton for the rest of his life. His personality is one of the most complicated, from a psychological point of view. Psychoanalysts have had a field day with him, and we must always keep in mind that he combined one of the supreme intellects of the human race with weaknesses that are hard to explain. His behavior can only be accounted for by considering that he differed from common mortals just as much emotionally as he did intellectually. Thus, it is very difficult to form an opinion of the man. There also were so many adulators during his life, as there are among his biographers, that one becomes skeptical of their utterances, especially when confronted with a certain number of documents and proven facts that cast dark shadows on Newton's personality.

A modern photograph of Trinity College, Cambridge. The windows of Newton's rooms are above the bush and to the left of the gate. Maxwell lived in nearby rooms around 1852. (Photograph courtesy of Sue Whytock.)

The Great Court of Trinity College, showing the Wren Library. Maxwell's rooms are at the left of the picture. (Photograph courtesy of Sue Whytock.)

Newton was a good student in elementary school but he did not show his extraordinary qualities early, except that he was very adept at building mechanical toys, some quite remarkable. His mother thought of making a farmer of him, since they had enough property that required attention and work. However, Newton did not show much inclination toward becoming a farmer, and some clergyman, either in the family or connected with the school he attended, persuaded his mother to let him attend Cambridge University. He was admitted as a student at Trinity College in 1661. Because he was not rich, he had to perform some menial services. He studied the normal curriculum, but there are records of his vast readings in optics, astronomy, mathematics, dynamics, chemistry, and alchemy. Furthermore, he had deeply studied the Bible, which he knew better than many theologians.

Newton formed his religious ideas in his early youth. He was nominally Anglican, but morally near to the Puritans for his austerity, discipline, and inclination to a sense of guilt. One finds many written self-examinations of his sins and thoughts. He consistently refused to take holy orders and be ordained, and he secretly held unorthodox opinions concerning the Trinity; in fact, he was Unitarian. This was a jealously guarded secret, but it is revealed in his private papers. Religious thought was a major occupation for Newton during his entire life, and there are voluminous writings on the subject among the papers he left.

He studied the Latin classics and Euclid's *Geometry* which at first he did not appreciate. He also studied Descartes's geometry, the first book on what today is called analytic geometry, and Wallis' *Arithmetica infinitorum,* one of the books that was to lead to the calculus by using infinite processes.

We must stress that Newton brought to fruition many ideas that were in the air. Special problems of what we call today differential and integral calculus had been solved previously many times, and infinite series had been used. Of the immediate predecessors of Newton in mathematics, Bonaventura Cavalieri (1598–1647), René Descartes (1596–1650), Pierre Fermat (1601–1665), Christiaan Huygens (1629–1695), John Wallis (1616–1703), James Gregory (1638–1675), and several others had each either solved particular problems or tried to create general theories, but none had succeeded in discovering the fundamental unifying ideas. Not even Newton saw them completely clearly, and many of his definitions or proofs would not be acceptable today, but the difference between the pre-Newton and the post-Newton era is colossal. He really opened the door, and the best proof of his success is the rush with which his immediate successors invaded the area he had opened and obtained a tremendous number of results that had been hiding at least since the time of the Greeks. That the time was ripe is shown by the independent discovery of the calculus by Gottfried Wilhelm Leibnitz (1646–1716), about two years after Newton's work. As we shall see, that work gave rise to a regrettable, undignified quarrel, but Leibnitz is reported to have said, "Taking mathematics from the beginning of the world to the time of Newton, what he has done is much the better half." And Newton, in the first edition of his *Philosophiae naturalis principia mathematica,* had acknowledged that when he revealed his methods, which he had kept secret, to Leibnitz, "that most distinguished man wrote back that he had also fallen upon a method of the same kind, and communicated his method which hardly differed from mine, except in his form of words and symbols." This passage was changed in later editions and does not appear in the English translation.

Newton was very jealous of his privacy, sometimes verging on the pathological. His secretiveness is reflected throughout his life in his publications and his utterances. The disclosure of Newton's scientific results often occurred in a peculiar way, strange even for his time, in which publication and communication methods were different from those of today. Newton often revealed his discoveries orally to some friend, but we know comparatively little about this. He also wrote private letters to friends or such people as Henry Oldenburg (1618–1677), the secretary of the Royal Society, who semiprofessionally transmitted what he received from others to a certain number of selected correspondents. Lastly, often years or decades later, Newton's results appeared in print either as communications to a learned society, such as the Royal Society, or in books, occasionally as appendices to unrelated books. Thus, as an example, a most important mathematical treatise, *De analysi per aequationes numero terminorum infinitas,* which contains fundamental results, was given to Isaac Barrow (1630–1670) in 1666, and Barrow communicated it three years later to John Collins (1625–1683) and Lord William Brouncker (1620–1684). It appeared in print in 1712. The *Opticks* lectures that were given in Cambridge around 1670 were sent to the secretary of the Royal Society in 1675 and appeared with many changes as a book in 1704. There is little doubt that such methods of publication contributed to the acrimonious priority arguments into which Newton entered at various times.

Newton was a prolific writer who left boxes full of his private papers. Those boxes have had a complicated history. Newton's niece and housekeeper, Catherine Barton, came into possession of them and left them to her daughter, the Countess of Portsmouth. The mathematical part of the so-called Portsmouth Collection was given in 1888 to the library of Cambridge University. The rest—a very large collection—was auctioned in 1936. The famous economist Lord Keynes acquired a substantial portion of the Portsmouth Collection and left it to Cambridge University. Some historians have had access to the papers and have perused them. They are now published in part in an immense editorial enterprise, but even so, they are understandable only with considerable effort. To comprehend them one needs to be a historian and a mathematician, and, of course, if one wants to delve into other aspects of Newton's thought, one has to also know the contemporary state of alchemy, religious ideas, astronomy, and so on. Little wonder that this man, so important for the history of human thought, is still in many ways an enigma. Lord Keynes is among those who have studied, at least in part, the Portsmouth Collection. The opinion of such a shrewd man, with extensive human knowledge and experience, carries weight. Some of his conclusions are reported in an address composed on the occasion of Newton's tercentenary, from which the following are excerpts:

> In the eighteenth century and since, Newton came to be thought of as the first and greatest of the modern age of scientists, a rationalist, one who taught us to think on the lines of cold and untinctured reason.
>
> I do not see him in this light. I do not think that any one who has pored over the contents of that box which he packed up when he finally left Cambridge in 1696 and which, though partly dispersed, have come down to us, can see him like that. Newton was not the first of the age of reason. He was the last of the magicians, the last of the Babylonians and Sumerians, the last great mind which looked out on the visible and intellectual world with the same eyes as those who began our intellectual inheritance rather less than 10,000 years ago. Isaac Newton, a posthumous child born with no father on Christmas Day, 1642, was the last wonderchild to whom the Magi could do sincere and appropriate homage. . . .
>
> In vulgar modern terms Newton was profoundly neurotic of a not unfamiliar type, but—I should say from the records—a most extreme example. His deepest instincts were occult, esoteric, semantic—with profound shrinking from the world, a paralyzing fear of exposing his thoughts, his beliefs, his discoveries in all nakedness to the inspection and criticism of the world. "Of the most fearful, cautious and suspicious temper that I ever knew," said Whiston, his successor in the Lucasian Chair. . . .
>
> His peculiar gift was the power of holding continuously in his mind a purely mental problem until he had seen straight through it. I fancy his pre-eminence is due to his muscles of intuition being the strongest and most enduring with which a man has ever been gifted. Anyone who has ever attempted pure scientific or philosophical thought knows how one can hold a problem momentarily in one's mind and apply all one's powers of concentration to piercing through it, and how it will dissolve and escape and

you find that what you are surveying is a blank. I believe that Newton could hold a problem in his mind for hours and days and weeks until it surrendered to him its secret. Then being a supreme mathematical technician he could dress it up, how you will, for purposes of exposition, but it was his intuition which was pre-eminently extraordinary—"so happy in his conjectures," said de Morgan, "as to seem to know more than he could possibly have any means of proving." The proofs, for what they are worth, were, as I have said, dressed up afterwards—they were not the instrument of discovery.

[From J. R. Newman, *The World of Mathematics* (New York: Simon & Schuster, 1956).]

Revelations to the Cambridge Student

At Cambridge the student Newton clearly made an impression on his mathematics proessor, Isaac Barrow. Barrow was a creative, expert mathematician who had had a very adventurous life, including travel in faraway countries, such as Turkey. Barrow knew Greek, Latin, and Hebrew well and some Arabic. He was an extrovert, an eloquent preacher who became King's chaplain. At the time of Newton's arrival at Cambridge, he was Lucasian Professor of Mathematics; however, he had been also Regius Professor of Greek and had taught optics as Lucasian Professor. Barrow had an unusually rich library, which contained scientific and theological works, including some of heretical philosophers, such as Spinoza and Hobbes, and he let his pupil see them.

Isaac Barrow (1630–1677), a protector and teacher of the young Newton and possibly one of his very few friends. Newton succeeded Barrow as Lucasian Professor at Cambridge. (Courtesy of the National Portrait Gallery, London.)

Isaac Newton at age forty-six, by Sir Godfrey Kneller. This is the earliest portrait of Newton. (Courtesy of Lord Portsmouth and the Trustees of the Portsmouth Estate.)

In 1665 Newton earned the degree of bachelor of arts, and was ready to embark upon creating new mathematics. In the same year, the plague struck England, and the university was closed to avoid contagion. Newton then left Cambridge and returned home and to his mother, always a strong influence in his life. At Woolsthorpe he received what one could call divine inspiration, and in a short period, he made the most important of the discoveries that were to make him immortal. He discovered the binomial theorem, the method for calculating the quantity $(1 + x)^p$ where p is any number. The formula for $p = n$, a positive integer, is simple; it was known before Newton's time, and can be obtained by careful reckoning and repeated multiplication. The result is a polynomial:

$$(1 + x)^n = 1 + \frac{n}{1} x + \frac{n(n-1)}{1 \cdot 2} x^2$$
$$+ \frac{n(n-1)(n-2)}{1 \cdot 2 \cdot 3} x^3 + \frac{n(n-1)(n-2) \cdots 1}{1 \cdot 2 \cdot 3 \cdots n} x^n.$$

If p is not a positive integer, the situation is much more complicated. We could try to extend formally the result, writing again

$$(1 + x)^p = 1 + \frac{p}{1} x + \frac{p(p-1)}{1 \cdot 2} x^2$$
$$+ \cdots \frac{p(p-1)(p-2) \cdots (p-n+1)}{1 \cdot 2 \cdot 3 \cdots n} x^n + \cdots.$$

However, the polynomial does not terminate; it becomes a series with an infinite number of terms that for $|x| < 1$ become smaller and smaller, and the sum of the series tends to a limit equal to $(1 + x)^p$. This example opens the door to the use of infinite series in analysis, and it gave Newton an object on which to develop the concepts of limit, infinitely small quantity, and, ultimately, the calculus. Newton's arguments would not be acceptable today, but they gave him the key. That things that today appear simple were in fact quite complicated is shown, even in the matter of notation, by the way in which Newton formulates the binomial theorem.

He sent the following letter to Oldenburg on June 13, 1676, for transmission to Leibnitz and then reproduced it many years later, in 1712, when he was embroiled in his quarrels with Leibnitz:

Although the modesty of Dr. Leibnitz in the Excerpts which you recently sent me from his Letter, attributes much to my work in certain Speculations regarding *Infinite Series*, rumor of which is already beginning to spread, I have no doubt that he has found not only a method of reducing any Quantities whatsoever into Series of this type, *as he himself asserts*, but also that he has found various Compendia, similar to ours if not even better.

Since, however, he may wish to know the discoveries that have been made in this direction by the English (I myself fell into this Speculation some years ago) and in order to satisfy his wishes to some degree at least, I have sent you certain of the points which have occurred to me.

Fractions may be reduced to Infinite Series by Division, and Radical Quantities may be so reduced by the Extraction of Roots. These Operations may be extended to Species in the same way as that in which they apply to Decimal Numbers. These are the Foundations of the Reductions.

The Extractions of Roots are much shortened by the Theorem

$$\overline{P + PQ}\bigg|^{\frac{m}{n}} = P^{\frac{m}{n}} + \frac{m}{n} AQ + \frac{m-n}{2n} BQ + \frac{m-2n}{3n} CQ + \frac{m-3n}{4n} DQ + \&c.$$

where $P + PQ$ stands for a Quantity whose Root or Power or whose Root of a Power is to be found, P being the first Term of that quantity, Q being the remaining terms divided by the first term, and $\frac{m}{n}$ the numerical Index of the powers of $P + PQ$. This may be a Whole Number or (so to speak) a Broken Number; a positive number or a negative one. For, as the Analysts write a^2 and a^3 &c. for aa and aaa, so for $\sqrt{a}, \sqrt{a^3}, \sqrt{c.a^5}$, &c. I write $a^{1/2}, a^{3/2}, a^{5/3}$, &c.; for $\frac{1}{a}, \frac{1}{aa}, \frac{1}{aaa}, a^{-1}, a^{-2}, a^{-3}$; for

$\frac{aa}{\sqrt{c.a^3 + bbx}}$, $aa \times \overline{a^3 + bbx}\big|^{-1/3}$; and for $\frac{aab}{\sqrt{c:a^3 + bbx \times a^3 + bbx:}}$, I write

$aab \times \overline{a^3 + bbx}\big|^{-2/3}$. In this last case, if $\overline{a^3 + bbx}\big|^{-2/3}$ be taken to mean $P + PQ$ in the Formula, then will $P = a^3, Q = bbx/a^3, m = -2, n = 3$. Finally, in place of the terms that occur in the course of the work in the Quotient, I shall use A, B, C, D, &c.

Thus A stands for the first term $P^{m/n}$; B for the second term $\frac{m}{n} AQ$; and so on. The use of this Formula will become clear through Examples.

[*Commercium Epistolicum,* 1712 edition. Translated by Eva M. Sanford, in W. F. Magie, *A Source book in Mathematics* (Cambridge, Mass.: Harvard University Press, 1935).]

This was probably the first of the great Newtonian discoveries; it may have occurred while he was still at Cambridge. We do not know because Newton did not publish it (and much more that was to follow) for many years.

At Woolsthorpe he also discovered the law of gravitation and the dynamics of the solar system. The story of the apple, according to which the thought of universal gravitation occurred to him when an apple fell from a tree under which he was resting, may or may not be true, but there are the words of the old Newton, recollecting that time:

In the same year [1666] I began to think of gravity extending to the orb of the moon, and having found out how to estimate the force with which a globe revolving within a sphere presses the surface of the sphere, from Kepler's rule of the periodical times of the planets being in a sesquialterate proportion of their distances from the centres of their orbs I deduced that the forces which keep the planets in their orbs must be reciprocally as the squares of their distances from the centres about which they revolve: and thereby compared the force requisite to keep the moon in her orb with the force of gravity at the surface of the earth, and found them answer pretty nearly. All this was in the two plague years of 1665 and 1666, for in those days I was in the prime of my age for invention, and minded mathematics and philosophy more than at any time since.

What Mr. Huygens has published since about the centrifugal forces I suppose he had before me. At length in the Winter between 1676 and 1677 I found the proposition that by a centrifugal force reciprocally as the square of the distance a planet must revolve in an ellipsis about the center of the force placed in the lower ombilicus of the ellipsis and with a radius drawn to that center described areas proportional to the time. . . .

It is clear that in order to be able to make the calculations mentioned above, Newton must have had several notions concerning the calculus and the fundamental laws of mechanics, too. For the first, we are again in the dark on the precise date of the invention. The passage from the binomial theorem to the concept of derivative and integral is not simple, but there are connections, and Newton somehow must have found them.

For the analysis of the dynamics of the universe, the explanation of the motion of the planets and universal gravitation, he must have used his own discoveries, but the detailed sequence of events is unknown. Only Newton could have informed us and he chose to remain silent. To unravel the planetary motion, the mathematical tool was not sufficient. He had to know the laws of mechanics, too. A version of the law of inertia was known to Galileo. Examples of the other laws had been found and applied in particular cases, for instance, by Galileo, Descartes, and Huygens, but there was no explicit general formulation, and to apply the mechanical laws derived from limited terrestrial experience to the universe required boldness, power of abstraction, and technical proficiency of the highest order.

Finally, in this extraordinary period, Newton also formed a theory of light supported by his own experiments that elucidated many optical phenomena, above all, the relation of the colors of the spectrum to white light, and established an objective relation between the physics of light and the physiology of our visual perceptions. The principles of Newton's optics have survived only in part, but in their development he uses so many ideas, methodological principles, and experimental virtuosity that we must count them as one of his major achievements.

Newton did not publish any of his great discoveries at the time he made them, but he revealed part of them to his mentor, Barrow, when he returned to Cambridge at the end of the plague. Following what was to become a behavioral pattern, he kept secret most of his results and published them only when prodded by somebody or when others found the same things. He became upset if his priority was not immediately acknowledged, or if somebody questioned his results. Why did he keep his discoveries secret? A possible partial explanation might be that he wanted to preserve a sort of monopoly on his fields of research; he certainly abhorred competition and arguments, and he may have wanted to have time to exploit his findings. That may be at most a partial explanation, and possibly not the most cogent.

The young Newton insistently professed indifference to fame and abhorrence of controversy. For example, in a letter to Collins in 1669, he wrote:

> That solution of the annuity problem if it will be of any use, you have my leave to insert it into the Philosophical Transactions, so it be without my name to it. *For I see not what there is desirable in public esteem, were I able to acquire and maintain it. It would perhaps increase my acquaintance, the thing which I chiefly study to decline.* Of that problem I could give exacter solutions, but that I have no leisure at present for computations.

And in another letter, to Oldenburg in 1676, he stated:

> . . . I beg your patience a week longer, I see I have myself a slave to philosophy, but if I get free of Mr. Linus's business, I will resolutely bid adieu to it eternally, excepting what I do for my private satisfaction, or leave to come out after me; for I see a man must either resolve to put out nothing new, or to become a slave to defend it.

The extrovert Barrow obtained permission from Newton to communicate some of his results to Collins, a member of the Royal Society at London who entertained lively correspondence with many colleagues. Newton at first had forbidden Barrow to reveal the name of the author of the mathematical papers, but later he relented, and in a letter the proud Barrow says to Collins, "I am glad my friend's paper gives you so much satisfaction; his name is Mr. Newton, a fellow of our college and very young (being but the second youngest Master of Arts) but of an extraordinary genius and proficiency in these things."

In 1669 Barrow resigned as Lucasian Professor of Mathematics. He may have done so to make way for Newton, whose genius he had undoubtedly recognized, and he certainly influenced the choice of his successor. However, by 1673 Barrow was back in Cambridge as Master of Trinity College, and Newton, with unusual warmth, wrote to Collins, "We shall enjoy Dr. Barrow again, especially in the circumstances of Master, nor does any rejoice at it more than, Sir, Your obliged humble Servant, I. Newton."

Lucasian Professor: Light Decomposed

The Lucasian professorship gave Newton £100 per year. Other income doubled this amount, which solved any economic problems he might have had. From then on for the rest of his life, Newton was prosperous, and at his death he left an estate of £36,000, which at the time was a very substantial sum. Newton had simple but expensive tastes, and he was quite generous with younger people and especially with members of his family.

In 1672 Newton was elected a member of the Royal Society. The society had been founded in 1660 by twelve private individuals interested in scientific investigations. It was part of the movement of scientific renaissance that brought about in 1603 the founding in Rome of the Accademia dei Lincei, of which Galileo had been the most distinguished member, and that was to bring in existence the Académie des Sciences in Paris (1666) and the Berlin academy, planned by Leibnitz in 1700. Those societies were initially supported by their members, met periodically, and demonstrated interesting experiments at the meetings. They also published their transactions, held discussions, and in general had an active function in the fostering of science. The Royal Society is possibly the one with the most impressive record of all. King Charles II gave it its royal charter in 1662, and it has remained continuously active ever since. In the early years, the secretaries of the society were often important in maintaining communications with the scientific world abroad, and one favorite method of communication was to write to the secretary of a society, asking him to pass information along to other interested persons. Oldenburg and Robert Hooke (1635–1702) were among the early secretaries of the Royal Society and were diligent in fulfilling their obligations. The president of the society was often more a patron of the sciences than a scientist. When Newton became president of the Royal Society in 1703, he had among his predecessors Samuel Pepys, the famous diarist, who was a high civil servant, but not a scientist.

In 1662 at Cambridge, Newton had already started experimenting with light. No organized physics laboratories existed at the time, and Newton performed his experiments in his rooms at Trinity College with apparatus built with his own hands. There is abundant evidence of Newton's excellent manual ability. One of his first projects was to build a reflecting telescope. For it he prepared the alloy of the mirror and ground it. We know also that he ground lenses and built the grinding machine. Before making the reflecting telescope he had experimented with more

The telescope Newton presented to the Royal Society in London. He lectured on telescopes in 1672 at the Society, shortly before being made a member. (Royal Society, London).

conventional refracting instruments. In order to improve the instrument's performance, he wanted to get rid of the colors that appear at the boundary of images, what is called chromatic aberration in modern language. To devise a method of attack to the problem, he passed ordinary white solar light through a prism and decomposed it into light of many colors. From this he was drawn into the general study of optics. The laws of reflection and refraction were known to him from previous investigators. He noted that lights of different colors have different refractive indexes, but he did not perceive that the dispersion of different glasses might be different. He thus arrived at the erroneous conclusion that it was impossible to make an achromatic lens (a lens that would not have chromatic aberration) by combining two lenses of different glass, one convergent and one divergent. He decided that to avoid chromatic aberration, it was necessary to avoid lenses completely and to build the telescope using only reflecting mirrors. He invented and built such a telescope. The instrument performed satisfactorily and it elicited the admiration of the members of the Royal Society, especially its secretary, Oldenburg. Newton presented it to the society where it is still preserved.

It is difficult to recreate the state of mind with which Newton must have studied optics. Very little was known about light at the time, and some things that now are self-evident then posed major problems. For instance, in optics there is a complicated mix between our sensory experience and the physics of light. Colors are definitely a subjective experience, but what is their physical basis? Are colors created by the eye? By the brain? Or are they objective qualities of light? And what is the relation of white, brown, and similar colors to the colors obtained by a prism? Those were some of the preliminary questions Newton had to answer.

Newton's first scientific paper, published in 1672 in the *Philosophical Transactions of the Royal Society* was on optics. In it he connected refrangibility, or refractive index, as we would say now, with color. The paper was assailed by some members of the society, including Hooke, a distinguished scientist whose name is forever connected with Hooke's law on the elastic behavior of solids. Newton's reaction was withdrawal and a resolution to refrain from publishing of further studies. Newton wrote to Oldenburg: "I intend to be no farther solicitous about matters of Philosophy: and therefore I hope you will not take it ill if you find me never doing anything more in that kind."

That, however, was an exaggeration. When he became Lucasian Professor, he gave courses in optics, which seem to have been attended by very few students. He did not publish his work for some time, as usual. About the same period, important discoveries in optics were made by Gregory, who also described a reflecting telescope, by the Jesuit Father Francesco Grimaldi (1618–1663), who discovered the phenomenon of diffraction (1665), by Hooke, who discovered the colors of thin plates (1666), and by Erasmus Bartholin (1625–1698), who discovered double refraction (1670).

Newton left his early lectures on optics in the archives of Cambridge University. They were copied and circulated, but were published only in 1729, after Newton's death. In the meantime, Newton incorporated his work in optics into a book that appeared in 1704. The "Advertisement" preceding the book gives an impression of Newton's attitude toward publication and of his complicated personality.

An elucidation on biblical archaeology by Newton in his own hand. (Courtesy of the Humanities Research Center, University of Texas at Austin.)

Part of the ensuing Discourse about Light was written at the Desire of some Gentlemen of the *Royal-Society,* in the Year 1675, and then sent to their Secretary, and read at their Meetings, and the rest was added about twelve Years after to complete the Theory; except the third Book, and the last Proposition of the Second, which were since put together out of scatter'd Papers. To avoid being engaged in Disputes about these Matters, I have hitherto delayed the printing, and should still have delayed it, had not the Importunity of Friends prevailed upon me. If any other Papers writ on this Subject are got out of my Hands they are imperfect, and were perhaps written before I had tried all the Experiments here set down, and fully satisfied my self about the Laws of Refractions and Composition of Colours. I have here publish'd what I think proper to come abroad, wishing that it may not be translated into another Language without my Consent.

The Crowns of Colours, which sometimes appear about the Sun and Moon, I have endeavoured to give an Account of; but for want of sufficient Observations leave that Matter to be farther examined. The Subject of the Third Book I have also left imperfect, not having tried all the Experiments which I intended when I was about these Matters, nor repeated some of those which I did try, until I had satisfied my self about all their Circumstances. To communicate what I have tried, and leave the rest to others for farther Enquiry, is all my Design in publishing these Papers.

In a Letter written to Mr. *Leibnitz* in the year *1679,* and published by Dr. *Wallis,* I mention'd a Method by which I had found some general Theorems about squaring Curvilinear Figures, or comparing them with the Conic Sections, or other the simplest Figures with which they may be compared. And some Years ago I lent out a Manuscript containing such Theorems, and having since met with some Things copied out of it, I have on this Occasion made it publick, prefixing to it an *Introduction,* and subjoining a *Scholium* concerning that Method. And I have joined with it another small Tract concerning the Curvilinear Figures of the Second Kind, which was also written many Years ago, and made known to some Friends, who have solicited the making it publick.

I. N.

April 1, 1704.

Note the remote connection of the last paragraph to the subject of optics, but its relation to Newton's quarrels.

About two hundred years later, when one thought optics was really understood, Einstein was once more confronted with fundamental problems concerning the nature of light, and he had to revolutionize the ideas prevailing in the early years of the twentieth century. His testimony is thus significant indeed, and here are his words prefacing Newton's *Opticks:*

Fortunate Newton, happy childhood of science! He who has time and tranquillity can by reading this book live again the wonderful events which the great Newton experienced in his young days. Nature to him was an open book, whose letters he could read without effort. The conceptions which he used to reduce the material of experience to order seemed to flow spontaneously from experience itself, from the beautiful experiments which he ranged in order like playthings and describes with an affectionate wealth of detail. In one person he combined the experimenter, the theorist, the mechanic and, not least, the artist in exposition. He stands before us strong, certain, and alone: his joy in creation and his minute precision are evident in every word and in every figure.

Reflexion, refraction, the formation of images by lenses, the mode of operation of the eye, the spectral decomposition and the recomposition of the different kinds of light, the invention of the reflecting telescope, the first foundations of colour theory, the elementary theory of the rainbow pass by us in procession, and finally come his observations of the colours of thin films as the origin of the next great theoretical advance, which had to wait, over hundred years, the coming of Thomas Young.

Newton's optics is not a complete science; it is science in the making. Often one describes it as a corpuscular theory. To a certain extent it is, but Newton mixes corpuscular concepts with others, such as "fits of easy reflections" and "fits of easy transmission," which are not easily classified. I believe he could very easily have accommodated himself to modern theories in which light has a double nature, wavelike and corpuscular at the same time.

The queries at the end of Newton's *Opticks* afford us a glimpse of his method of thinking and working. They show his very extensive knowledge of chemical and technological properties of materials, and, at the same time, his powerful ability to observe and, when possible, measure. To appreciate them, we must always remember that Newton is navigating uncharted seas and has very little precedent to follow. His observations resemble those of an extraordinarily intelligent child left to his own devices. From his observations, his great mind abstracts, although incompletely, what later was accepted as the truth. Once in a while Newton recognizes deep analogies between light and waves, and we think that he is about to develop a wave theory of light, but no; other phenomena, equally important, come to his mind and he turns corpuscular. Soon after, that which we could call properties

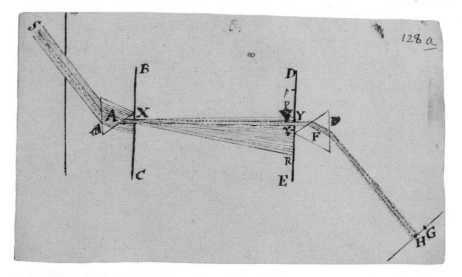

A diagram by Newton of the fundamental experiment connecting color with refractive index. The first prism acts as a monochromator, in modern language. The second serves to measure the refractive index of each color. (By permission of the Syndics of Cambridge University Library.)

of the black body attract his attention, and he makes shrewd observations on the subject. If we wanted to read Newton in such a way as to make him a sire of all modern physics, it would be possible. I do not believe it would be fair. He is a great mind bewildered by what he observes. He has a powerful analytical ability and critical sense, and thus he can recognize what he can encompass in a theory and what escapes him. Repeatedly, he expresses his diffidence for hypotheses and he clearly distinguishes what seems well established and what is speculation.

His *Opticks* starts with the declaration: "My design in this book is not to explain the properties of light by hypotheses, but to propose and prove them by reason and experiments. . . ." This is one of the many times when Newton objects to hypotheses. We should inquire into what he means by that. From his own practice we find that he did frame hypotheses many times. However, he required that known experimental facts be compatible with them, and that the hypothesis be verifiable. If an hypothesis could not be tested by submitting its consequences to experiments, it should be viewed with suspicion. Furthermore, one should keep a clear distinction between hypothesis and verified facts. Among the most important hypotheses developed by Newton are gravity, the nature of white light, the corpuscular nature of light, and the hypothesis of "fits" of light.

The *Principia* and the Fabric of the Universe

The unraveling of the dynamics of planetary action and the discovery of universal gravitation are the achievements of Newton that have most impressed the popular mind. We have seen the line of thought of the young Newton at Woolsthorpe. Considering the moon on a circular orbit around the earth, he found that the centrifugal force balanced gravity, if one extended gravity as measured on the earth, to the lunar orb, and assumed that it decreased by an inverse-square law. The calculation, in first approximation, is simple enough. It is said that it did not give a good numerical agreement because the value of the radius of the earth used by Newton was not correct.

There is another inherent difficulty. The calculation is simple if one neglects the sizes of the earth and the moon and considers their masses as if they were concentrated at their centers. A remarkable theorem proved by Newton himself, but years later, shows that for an inverse-square attraction law this simplification is exactly permissible. Thus, there were two flaws in Newton's discovery. Both were later removed — one by a better measurement of the terrestrial radius, the other by Newton's own discovery. Perhaps those were the objective reasons that restrained Newton from publishing the result, but, given his personality, no special objective reasons are needed to explain his silence.

In the meantime, the inverse-square law occurred to several other persons. It is not such a farfetched hypothesis as to require a Newton for its surmise. Hooke, Sir Christopher Wren, the architect of St. Paul's, and the well-known astronomer Edmund Halley (1656–1742) talked about it in January 1684. Halley described the meeting as follows:

I met with Sir Christopher Wren and Mr. Hooke, and falling in discourse about it, Mr. Hooke affirmed, that upon that principle all the laws of the celestial motions were to be demonstrated, and that he himself had done it. I declared that ill success of my own attempts; and Sir Christopher, to encourage the enquiry, said, that he would give Mr. Hooke, or me, two months time, to bring him a convincing demonstration thereof; and besides the honour, he of us, that did it, should have from him a present of a book of 40 shillings. Mr. Hooke then said, that he had it, but he would conceal it for some time, that others trying and failing might know how to value it, when he should make it public. However I remember, that Sir Christopher was little satisfied that he could do it; and though Mr. Hooke then promised to show it to him, I do not find, that in that particular he has been so good as his word.

Time passed and no solution arrived. Halley then thought to go to Cambridge and consult Newton on the subject. When he asked him what would be the trajectory of a body attracted to a fixed point by an inverse-square law, Newton answered immediately, to Halley's pleased surprise: "An ellipse, with the attracting center in the focus." Halley asked how he knew it, and Newton answered, "I have calculated it." However, he could not find the calculation, but promised Halley to send it to him within a few weeks, which he did, even giving two proofs of the theorem. In fact, he had made the calculation several years earlier, but he had kept the result to himself, according to what he wrote in 1714. The result is remarkable and certainly not obvious. At first it seems strange that a central symmetrical force should produce an elliptical orbit, although a closer consideration shows that no symmetry law is violated because the initial conditions may be asymmetric. The calculation, even by modern means, is sophisticated.

At the end of 1684, Newton sent Halley a small treatise, *De motu corporum* ("On the Motion of Bodies"), which contained the principles of mechanics and was based on Newton's lectures. Halley recognized that it was a most important work and decided to try to convince Newton to reveal his discoveries in full.

Halley was a man of the world, an astute psychologist and a great astronomer. It is not known how he persuaded Newton to publish his work, but with the support of the Royal Society he succeeded, and by the end of 1685 Newton was at work on the first book of the *Philosophiae naturalis principia mathematica*, probably the greatest scientific book ever written. Within eighteen months the work was completed. The original is in Latin. In a modern English translation of the third edition of 1725, it is 547 pages long and divided in three parts. First, there is an ode to Newton by Halley, then a preface by Newton from the first edition, followed by one by Roger Cotes, who helped prepare the second edition. The book is written in the style of Greek geometry, using geometrical proofs throughout. There is little doubt that many of the results have been obtained otherwise, using analytical methods either known to Newton's contemporaries or invented by him. The mathematical technique used does not facilitate the reading for a modern student, but perhaps geometrical methods were more familiar to Newton's contemporaries. He told a friend that "to avoid being bated by little smatterers in mathematics [he] designedly made [his] principle abstruse; but yet so as to be understood by able mathematicians."

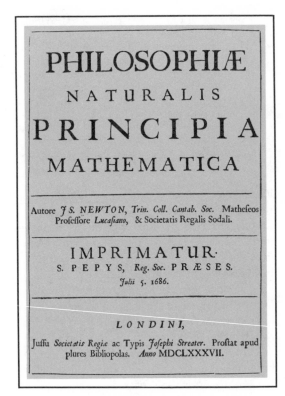

The title page of the first edition of the Principia *(1687). Editions published during Newton's lifetime differ slightly from each other, reflecting developments in the author's thoughts. (Courtesy of the Bancroft Library, University of California, Berkeley.)*

Before starting Book I, he introduces sundry definitions and axioms. We see him grappling here with some of the most profound and difficult problems of the foundations of mechanics. The definition of mass, "The quantity of matter is the measure of the same, arising from its density and bulk conjointly," if given by a lesser light than Newton, would have been deemed tautological, but that is what he wrote. He points out also "that the same {mass} is known by the weight of each body for it is proportional to the weight as I have found by experiments on pendulums, very accurately made. . . ." That last remark had to wait for Einstein to find a deep explanation. Newton must have meditated profoundly on the concepts of time and space. He recognized that they contained something that had to be analyzed. He arrived at definitions adequate for his purpose, but again the problem was ultimately carried to a much deeper understanding by Einstein. The axioms or laws of motion are three: (1) "Every body continues in its state of rest or of motion in a right line, unless it is compelled to change that state by forces impressed to it." (2) "The change of motion {momentum in modern language} is proportional to the motive force impressed; and is made in the direction of the right line in which that force is impressed." (3) "To every action there is always opposed an equal reaction; or, the mutual actions of two bodies upon each other are always equal, and directed to contrary parts."

Book I then follows. It begins with a mathematical section adumbrating the concept of limit and then gives a treatise on mechanics. In particular, it studies motion under central forces in general, specializing later to the inverse-square law. A geometrical study of conic sections, needed for further progress, is inserted. Attraction by extended bodies forms the subject of other sections, including also the theorem that is mentioned on page 60.

Book II studies motion in resisting mediums. Newton assumes a resistance proportional to the velocity or to the square of the velocity and investigates both cases. He also considers the combined action of a central force and a resistance. Density and compression of fluids and elementary hydrostatics follow. The action of the resistance of the medium on pendular motion comes next. The study of the influence of the shape of a body moving in a medium offers the occasion for new mathematics, what is now called calculus of variations. The last part of Book II is devoted to waves, with due credit to Huygens. Newton considers waves in water as well as in air. He calculates the velocity of propagation of sound, but the result is for an isothermal, not an adiabatic, compression. It thus disagrees with experiment. The book closes with a study of rotating fluids.

Book III is entitled "System of the World: In Mathematical Treatment." With the same principles demonstrated in the first two books, Newton now plans to "demonstrate the frame of the System of the World." The subject calls for an

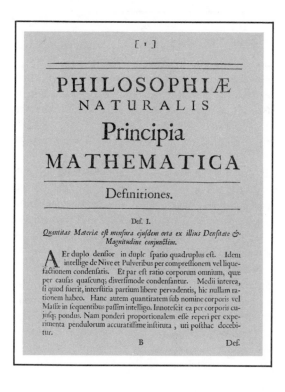

The beginning of Book I of the Principia. *(Courtesy of the Bancroft Library, University of California, Berkeley.)*

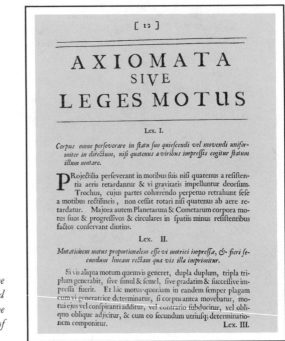

[12]

AXIOMATA
SIVE
LEGES MOTUS

Lex. I.

Corpus omne perseverare in statu suo quiescendi vel movendi uniformiter in directum, nisi quatenus a viribus impressis cogitur statum illum mutare.

PRojectilia perseverant in motibus suis nisi quatenus a resistentia aeris retardantur & vi gravitatis impelluntur deorsum. Trochus, cujus partes cohaerendo perpetuo retrahunt sese a motibus rectilineis, non cessat rotari nisi quatenus ab aere retardatur. Majora autem Planetarum & Cometarum corpora motus suos & progressivos & circulares in spatiis minus resistentibus factos conservant diutius.

Lex. II.

Mutationem motus proportionalem esse vi motrici impressae, & fieri secundum lineam rectam qua vis illa imprimitur.

Si vis aliqua motum quemvis generet, dupla duplum, tripla triplum generabit, sive simul & semel, sive gradatim & successive impressa fuerit. Et hic motus quoniam in eandem semper plagam cum vi generatrice determinatur, si corpus antea movebatur, motui ejus vel conspiranti additur, vel contrario subducitur, vel obliquo oblique adjicitur, & cum eo secundum utriusq; determinationem componitur.   Lex. III.

Two pages of Latin text of the "Laws of Motion," as formulated in the Principia. *(Courtesy of the Bancroft Library, University of California, Berkeley.)*

introduction on "Rules of reasoning in philosophy," in which Newton discusses some important epistemological questions and questions concerning the divisibility of matter. He then proceeds to a description of the astronomical phenomena as observed, that is, the motion of planets and satellites in the solar system, and he explains them in quantitative detail on the sole basis of universal gravitation. This is a colossal and mathematically difficult feat, based on the available observations of Flamsteed, Halley, and others.

A general scholium, or remark, ends the work. Newton fights the Cartesian hypothesis of vortices, which at that time was given great credit among natural philosophers. I will not delve into Cartesian physics. Descartes is considered the founder of modern philosophy, but he was also a mathematician and scientist. In philosophy and mathematics his work was of supreme importance. His physics should be studied by anyone who wants to deeply understand the development of physics, but its technical results are modest and removed from modern science.

The second part of the scholium deals with Newton's idea of God, and the third gives some of his scientific philosophy. Because of the importance of this text, excerpts are given below, omitting chiefly the theological part. In the next-to-last paragraph appears the famous statement, *"Hypotheses non fingo"* ("I frame no hypotheses.") Its true meaning will appear only by reading it in context.

[13]
Lex. III.

Actioni contrariam semper & æqualem esse reactionem: sive corporum duorum actiones in se mutuo semper esse æquales & in partes contrarias dirigi.

Quicquid premit vel trahit alterum, tantundem ab eo premitur vel trahitur. Siquis lapidem digito premit, premitur & hujus digitus a lapide. Si equus lapidem funi alligatum trahit, retrahetur etiam & equus æqualiter in lapidem: nam funis utrinq; distentus eodem relaxandi se conatu urgebit Equum versus lapidem; ac lapidem versus equum, tantumq; impediet progressum unius quantum promovet progressum alterius. Si corpus aliquod in corpus aliud impingens, motum ejus vi sua quomodocunq; mutaverit, idem quoque vicissim in motu proprio eandem mutationem in partem contrariam vi alterius (ob æqualitatem pressionis mutuæ) subibit. His actionibus æquales fiunt mutationes non velocitatum sed motuum, (scilicet in corporibus non aliunde impeditis :) Mutationes enim velocitatum, in contrarias itidem partes factæ, quia motus æqualiter mutantur, sunt corporibus reciproce proportionales.

Corol. I.

Corpus viribus conjunctis diagonalem parallelogrammi eodem tempore describeret, quo latera separatis.

Si corpus dato tempore, vi sola M, ferretur ab A ad B, & vi sola N, ab A ad C, compleatur parallelogrammum ABDC, & vi utraq; feretur id eodem tempore ab A ad D. Nam quoniam vis N agit secundum lineam AC ipsi BD parallelam, hæc vis nihil mutabit velocitatem accedendi ad lineam illam BD a vi altera genitam. Accedet igitur corpus eodem tempore ad lineam BD sive vis N imprimatur, sive non, atq; adeo in fine illius temporis reperietur alicubi in linea illa

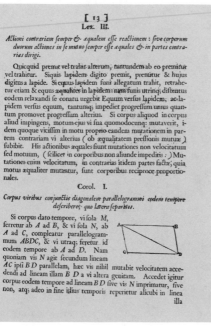

General Scholium

The hypothesis of vortices is pressed with many difficulties. That every planet by a radius drawn to the sun may describe areas proportional to the times of description, the periodic times of the several parts of the vortices should observe the square of their distances from the sun; but that the periodic times of the planets may obtain the $\frac{3}{2}$th power of their distances from the sun, the perodic times of the parts of the vortex ought to be as the $\frac{3}{2}$th power of their distances. That the smaller vortices may maintain their lesser revolutions about Saturn, Jupiter, and other planets, and swim quietly and undisturbed in the greater vortex of the sun, the periodic times of the parts of the sun's vortex should be equal; but the rotation of the sun and planets about their axes, which ought to correspond with the motions of their vortices, recede far from all these proportions. The motions of the comets are exceedingly regular, are governed by the same laws with the motions of the planets, and can by no means be accounted for by the hypothesis of vortices; for comets are carried with very eccentric motions through all parts of the heavens indifferently, with a freedom that is incompatible with the notion of a vortex.

Bodies projected in our air suffer no resistance but from the air. Withdraw the air, as is done in Mr. *Boyle's* vacuum, and the resistance ceases; for in this void a bit of fine down and a piece of solid gold descend with equal velocity. And the same argument must apply to the celestial spaces above the earth's atmosphere; in these spaces, where there is no air to resist their motions, all bodies will move with the greatest freedom; and the planets and comets will constantly pursue their revolutions in orbits given in kind and position, according to the laws above explained; but though these bodies may, indeed, continue in their orbits by the mere laws of gravity, yet they could by no means have at first derived the regular position of the orbits themselves from those laws.

From the order of the universe, he then argues:

This most beautiful system of the sun, planets, and comets, could only proceed from the counsel and dominion of an intelligent and powerful Being. And if the fixed stars are the centres of other like systems, these, being formed by the like wise counsel, must be all subject to the dominion of One; especially since the light of the fixed stars is of the same nature with the light of the sun, and from every system light passes into all the other systems: and lest the systems of the fixed stars should, by their gravity, fall on each other, he hath placed those systems at immense distances from one another.

This Being governs all things, not as the soul of the world, but as Lord over all; and on account of his dominion he is wont to be called *Lord God* παντοκράτωρ, or *Universal Ruler;* for *God* is a relative word, and has a respect to servants; and *Deity* is the dominion of God not over his own body, as those imagine who fancy God to be the soul of the world, but over servants. The Supreme God is a Being eternal, infinite, absolutely perfect; . . .

Hitherto we have explained the phenomena of the heavens and of our sea by the power of gravity, but have not yet assigned the cause of this power. This is certain, that it must proceed from a cause that penetrates to the very centres of the sun and planets, without suffering the least diminution of its force; that operates not according to the quantity of the surfaces of the particles upon which it acts (as mechanical causes used to do), but according to the quantity of the solid matter which they contain, and propagates its virtue on all sides to immense distances, decreasing always as the inverse square of the distances. Gravitation towards the sun is made up out of the gravitations towards the several particles of which the body of the sun is composed; and in receding from the sun decreases accurately as the inverse square of the distances as far as the orbit of Saturn, as evidently appears from the quiescence of the aphelion of the planets; nay, and even to the remotest aphelion of the comets, if those aphelions are also quiescent. But hitherto I have not been able to discover the cause of those properties of gravity from phenomena, and I frame no hypotheses [the famous *"Hypotheses non fingo"* in the Latin original]; for whatever is not deduced from the phenomena is to be called an hypothesis; and hypotheses, whether metaphysical or physical, whether of occult qualities or mechanical, have no place in experimental philosophy. In this philosophy particular propositions are inferred from the phenomena, and afterwards rendered general by induction. Thus it was that the impenetrability, the mobility, and the impulsive force of bodies, and the laws of motion and of gravitation, were discovered. And to us it is enough that gravity does really exist, and act according to the laws which we have explained, and abundantly serves to account for all the motions of the celestial bodies, and of our sea.

And now we might add something concerning a certain most subtle spirit which pervades and lies hid in all gross bodies; by the force and action of which spirit the particles of bodies attract one another at near distances, and cohere, if contiguous; and electric bodies operate to greater distances, as well repelling as attracting the neighboring corpuscles; and light is emitted, reflected, refracted, inflected, and heats bodies; and all sensation is excited, and the members of animal bodies move at the command of the will, namely, by the vibrations of this spirit, mutually propagated along the solid filaments of the nerves, from the outward organs of sense to the brain, and from the brain into the muscles. But these are things that cannot be explained in few words, nor are we furnished with that sufficiency of experiments which is required to an accurate determination and demonstration of the laws by which this electric and elastic spirit operates.

The text of the scholium, as well as other parts of the *Principia* differs in the various editions that appeared during Newton's life. Some of these changes reflect his changes of ideas, others pertain to the quarrels he had at the time with Hooke, John Flamsteed (1646–1719), and Leibnitz.

The Royal Society had urged Newton to compose the book, but in the end Halley had to subsidize the printing with his own money. It is strange that Newton, who certainly was not poor, did not contribute. The first edition sold quickly. Even during its composition, Newton had a dispute with Hooke, who had surmised the inverse-square law, but had not derived its consequences. Newton was very stingy in giving him any credit, in spite of Halley's entreaties. The book was reviewed with high praise in the *Acta eruditorum,* a leading German periodical, and in the *Journal des sçavans* in Paris. Not many, however, could really understand it. Even distinguished mathematicians, such as the young Abraham de Moivre (1667–1754), had serious difficulties. Leibnitz and Huygens were avid students of it, but, besides Newton, they were the best mathematicians and physicists in the world. David Gregory (1659–1708) in Scotland became an enthusiastic follower of the Newtonian doctrines and propagated them in the Scottish universities. Oxford followed, and lastly Cambridge, although Newton was there and taught the new mechanics. Later, the English were handicapped by a worship of Newton's published geometrical methods. The real developments that resulted from the *Principia* occurred more on the continent. Voltaire understood the importance of the work and popularized it in France after a visit to England in 1729, and from France the fashion for Newtonian doctrines spread all over cultivated Europe. For instance, in Italy, a well-known literary figure, Francesco Algarotti (1712–1764), wrote the book *Newtonianism for Ladies,* which became extremely popular. The fad reminds one of the fad for Einstein's relativity two hundred years later.

The true followers of Newton, who developed his ideas and propagated his views so that they became pervasive, were on the continent of Europe. The great heirs of the creators of calculus—the Bernoullis and Leonhard Euler (1707–1783)—and the French—Alexis Claude Clairaut (1713–1765) and Jean Le Rond d'Alembert (1717–1783)—used Leibnitz's notations and algorithms to carry on Newton's program. The *Principia* was recast in analytical form, and made accessible to a much wider circle than in the original geometrical vestments. More and more details of the system of the world and of the shape of the earth were explored. Newtonian mechanics was extended to a more complex system: fluids, strings, gases. It was found admirably precise and powerful in its predictive and explanatory ability.

The success of celestial mechanics was colossal. Not only did it predict celestial motions with a precision unknown before, but its fundamental ideas extended much beyond its original bounds, so much so that for a century the ideal of a scientific explanation was based on the Newtonian example, and it was thought that "to explain" was more or less equivalent to casting a certain family of phenomena into a Newtonian frame.

For two centuries Newtonian mechanics explained all details of planetary motion; some anomalies that seemed unaccountable were ascribed to the presence of a new planet. In 1846 Urbain Jean Joseph Le Verrier (1811–1877) and, independently, John Couch Adams (1819–1892) reached this conclusion and could even determine where the planet had to be—and there it was. Only the motion of the perihelion of Mercury showed a minute but untractable anomaly.

These impressive feats contributed to establish a profound faith in Newtonian mechanics and possibly retarded an examination of its foundations. It is likely that Newton suspected that not all was perfect there. But only in 1905, almost two hundred years after the publication of the *Principia,* did Einstein shake the basic concepts of time and space, and it was only in 1916 that Newton's gravitational theory was superseded by something that, although it gave almost identical numerical results, was conceptually radically different—general relativity. For a time, however, its advantages over Newton's gravitation, from an experimental point of view, were minute. The explanation of the motion of Mercury's perihelion was one of the greatest "practical" successes of general relativity.

Newton's achievements went far beyond the invention of the calculus, the development of mechanics, and his optical discoveries. In pure mathematics he left profound studies on the curves of the third order, numerical methods for the solution of equations, and much more that now has its place in the standard teaching of analysis or algebra.

In physics he studied thermometry, observing the fixed temperature at which water boils or freezes, the cooling law of hot bodies, and several other topics that pale only when compared with his major achievements.

Past the Prime of His Age for Invention

At the time of the appearance of the *Principia,* there was trouble between the King and the University of Cambridge. When James II tried to impose some Catholic professors on the university against its will, Newton was one of the delegates sent to London to defend the university's cause. Election to Parliament as a representative of the university followed in 1689.

As a member of Parliament, Newton had to travel to London more often than before. He became acquainted with Pepys and the illustrious philosopher John Locke, a most attractive character who became Newton's loyal friend. At that time Newton again gave some indications of his periodic disgust with science and indifference to the field in which he was supreme. He indicated to his friends that he would like to leave Cambridge life, and he put out feelers for various jobs, but with no success. Charles Montagu (1661–1715) the heir of an important English family who in 1699 became Lord Halifax, had befriended Newton in 1679, when he was a young student at Cambridge. Newton hoped to obtain a suitable position through his influence, and Montagu was trying to do his best to help his friend.

In 1689 Newton's mother died. He had been deeply attached to her; it was possibly the deepest emotional tie of his life, and his love was reciprocated by his mother. Newton had frequently visited her at Woolsthorpe from Cambridge, and when she became ill, he left London to nurse her. She left him a considerable legacy that assured Newton's economic affluence.

By 1692, however, the fifty-year-old Newton had started to display serious symptoms of psychological trouble. He fell into a great depression, complicated by delusions of persecution. His letters at the time show clearly his deranged state. He wrote of "being convinced that Mr. Montagu is false to me. I have done with him."

And on September 16, 1693, he wrote to Locke:

> Sir,
> Being of opinion that you endeavoured to embroil me with women and by other means, I was so much affected with it, as that when one told me you were sickly and would not live, I answered, 'twere better if you were dead. I desire you to forgive me this uncharitableness. For I am now satisfied that what you have done is just, and I beg your pardon for my having hard thoughts of you for it, and for representing that you struck at the root of morality, in a principle you laid down in your book of ideas, and designed to pursue in another book, and that I took you for a Hobbist. I beg your pardon also for saying or thinking that there was a design to sell me an office, or to embroil me.
> I am your most humble
And unfortunate servant,
> Is. Newton.

He wrote similar letters to Pepys and to others. The rumor spread, even beyond England, that Newton had lost his mind. His friends were greatly distressed and alarmed by his missives, and did not know how to answer.

There has been much speculation about Newton's mental illness, ranging from psychoanalytical interpretations to attribution to mercury poisoning. Fortunately, Newton recovered without any special care, and was well within about a year.

In the meantime, Montagu, who had generously disregarded the slurs of the sick Newton, succeeded in finding him a suitable post. He wrote to Newton in March 1696:

> I am glad that at last I can give you a good proof of my friendship, and the esteem the King has of your merits. . . . The King has promised to make Mr. Newton Warden of the Mint. The office is most proper to you. 'Tis the chief officer in the Mint. 'Tis worth five or six hundred pounds per annum, and has not too much business to require more attendance than you can spare. . . .

In fact, the office was number two in the mint, the chief officer being the Master, to which position Newton was elevated years later. Moreover, the business was far from a sinecure, at least for a certain period, because England was changing

The British one-pound bank note celebrates the most distinguished Master of the Mint. Note the proof of the elliptical planetary motion from the Principia, *the prism, the telescope, and the apple tree. (Reproduced with permission of the Bank of England.)*

its coinage, and this difficult and complicated operation required the most serious attention from the Warden of the Mint. The old coinage had been sheared by its users, resulting in coins containing less precious metal than legally expected. The government was in the process of retiring the old coins at their face value and issuing new ones with the rim of the coins milled in order to make it impossible to shear them.

In the first years of his stay in London, Newton was engrossed in the new coinage, and he did excellent work in accomplishing the difficult operation. His puritanical conscience made him unremitting in the prosecution of counterfeiters and totally adverse to possible compromises or bribes. His adversaries tried repeatedly, to no avail, to get rid of him by creating opportunities of more lucrative positions with less work.

As required by his new position, Newton established himself in London. At Cambridge, he had lived in Trinity College; in London, he had to set up house for himself. His half-sister had married a clergy named Barton and they had a daughter named Catherine, who at that time was sixteen years old. Catherine became Newton's housekeeper for the rest of his life. She was pretty, witty, and full of life, so much so that she soon became a society lady and contemporary chronicles speak extensively of her. Rumor, likely to be true, had it that she was the lover of Lord Halifax for years. In fact, his lordship bequeathed to her such substantial amounts that it was bound to generate rumors. On the other hand, it does not seem possible by the dates of the events that Catherine's graces may have influenced Newton's preferment for the mint. In 1715, after Halifax's death, Catherine married John Conduitt, who succeeded Newton at the mint and wrote one of his early biographies.

Once the recoinage operation was successfully finished, Newton was promoted to Master at the Mint in 1699. The office became less demanding and Newton had more leisure and better pay, but he did not resume his mathematical and physical work. He turned more and more to theology, alchemy, and chronological studies. Only a portion of his chronological work was published. The theological work was tainted with heresy, relating to questions on the Trinity, and that was sufficient reason to keep it sub-rosa. His alchemistic work was by its nature semimystical and secret. The following is from a letter dated April 26, 1676, from Newton to Oldenburg pertaining to it:

Sir,

Yesterday I reading the two last *Philosophical Transactions* had the opportunity to consider Mr. Boyle's uncommon experiment about the *incalescence of gold and mercury*. I believe the fingers of many will itch to be at the knowledge of the preparation of such a mercury; and for that end some will not be wanting to move for the publishing of it, by urging the good it may do in the world. But, in my simple judgement, the noble author, since he has thought fit to reveal himself so far, does prudently in being reserved in the rest. . . . Because the way by which mercury may be so impregnated, has been thought fit to be concealed by others that have known it, and therefore may possibly be an inlet to something more noble, not to be communicated without immense damage to the world, if there should be any verity in the Hermetic writers; therefore I question not, but that the great wisdom of the noble author will sway him to high silence, till he shall be resolved of what consequence the thing may be, either by his own experience, or the judgement of some other that thoroughly understands what he speaks about; that is, of a true Hermetic philosopher, whose judgement (if there be any such) would be more to be regarded in this point, than that of all the world beside to the contrary, there being other things beside the *transmutation of metals* (if those *great pretenders* brag not) which none but they understand. Sir, because the author seems desirous of the sense of others in this point, I have been so free as to shoot my bolt; but pray keep this letter private to yourself.

> Your servant,
>
> Isaac Newton.

More representative are the last "Queries" appended to his *Opticks*. They show a different man from the illuminated rationalistic author of the *Principia*.

It seems that Newton did not recognize the different value of his scientific and mathematical efforts from his other intellectual activities. Posterity remembers him only for the first, but he repeatedly professed disgust for them. Occasionally he expressed himself as if his scientific genius was almost outside of his own person and as if he was prey to it and would have liked to be liberated from it. The other occupations, however, must also have been engrossing. We have it from a witness, his copyist H. Newton (no relation), that at the time he was writing the *Principia* (an almost superhuman task, considering the speed of composition), he kept his alchemy studies going. He had a furnace lighted, and,

About 6 weeks at Spring and 6 at the Fall, the fire in the elaboratory [sic] scarcely went out, which was well furnished with chemical materials as bodies, receivers, heads, crucibles, etc., which was made very little use of, the crucibles excepted, in which he fused his metals; he would sometimes, though very seldom, look into an old mouldy book which lay in the elaboratory [sic]. I think it was entitled Agricola de Metallis, the transmuting of metals being his chief design, for which purpose antimony was a great ingredient. . . .

The alchemical studies of Newton are confined to his manuscripts; almost nothing was published. His genius did not extend to chemistry, and the founder of mathematical physics is not important in the history of chemistry. Possibly a different type of mind was needed to make progress in that science.

In mathematics he was one of the supreme inventors, but I doubt that most of his proofs would be considered acceptable today. Newton opened a vast new region, and he explored it without too much concern for details. He could do marvels with infinite series, but worries about convergence were not for him. What counts is the result, the power of solving new problems far beyond any previous reach. Because Newton was so self-centered, he did his mathematics for himself, without worrying too much about making it accessible to common mortals—hence, his disdain for notation and easy manipulation. That reflected on his methods and gave a great edge to Leibnitz's differential calculus over the mathematically equivalent Newtonian fluxions. The mathematical content of both theories is approximately the same, but Leibnitz's seed grew rapidly into an enormous tree, while Newton's languished.

There is also a tremendous amount of religious manuscripts by Newton, ranging from chronological and biblical studies to theological elucubrations. In religion a powerful mind such as Newton's is naturally prone to intellectual and theological studies, in contrast to simpler minds, which would probably stress the moral and behavioral teachings. Thus, to understand or appreciate Newton's work in the field of religion, one should be a theologian and a historian versed in the contemporary problems that he faced. All this is now past, and the interest is limited to the scholar; it is not that part of Newton's legacy which is still important and alive in modern science.

At this point I will digress for a short time on the religion of physical scientists. First, it is conditioned by the time in which they lived. Problems and feelings that may have been strong for Newton are almost incomprehensible today. Apart from this, childhood education and family tradition are possibly extremely strong factors. I have the impression, however, that science and religion are kept mostly in separate compartments of the brain, and thus we can find in eminent physicists all gradations, from deep ritual religion to avowed and proclaimed atheism. At one extreme I would put Newton and the mathematician Augustin-Louis Cauchy (1789–1857), one Protestant, one Catholic. In the middle ground there are the Victorians, such as Michael Faraday (1791–1867), who had a deep commitment to religion, although I believe not theologically rooted, or the simple but sincere religion of James Clerk Maxwell (1831–1879). Among the moderns, the Americans Robert Millikan (1868–1953) and Arthur Compton (1892–1962) relished preaching and seem to me as extremely simpleminded but fervent in religion. In the middle ground are Galileo and Einstein. The first made many pronouncements on religion, but the more one reads them, the more one feels that he spoke with tongue in cheek and that his religion was detached from formal Catholicism and turned into a reverence for Nature that would not satisfy any formal church. In changed times, and without the danger of being burned at the stake as a heretic, Einstein seems to me to be not far from Galileo, if one can guess at such deep and dim views and feelings as those connected with religion. Enrico Fermi (1901–1954) was outright agnostic, and at the extreme of avowed atheism stands Laplace, who stated that God, at least for astronomy, was "an unnecessary hypothesis." Religion, however, does not seem to have visibly influenced the scientific work of any of these scientists. When they discussed the relation of their discoveries to religion, I have the impression it was always ex post facto, and mostly when there were compelling external reasons for doing so.

Science Abandoned: Warden of the Mint and the Darker Side

It seems that while Newton was at the mint, he still had his old brain powers, because in 1696 he solved a new type of mathematical problem: the calculation of the line of fastest descent for a material heavy point. The problem had been posed by Johann Bernoulli (1667 – 1748) as a difficult challenge to all mathematicians and he had given six months for the solution. Newton heard of it one day after returning from his mint office and had the solution by the next morning. When Bernoulli saw it, without knowing its author, he is reported to have said, *"Tanquam ex ungue leonem"* ("From the claw you recognize the lion").

But Newton became more and more involved in many personal controversies with distinguished contemporaries. From reading the letters and documents, one possibly has an impression of more violent feelings than actually prevailed. The rhetoric of the seventeenth century is very different from that of the present, and just as we cannot literally take the "Your humble and devoted servant" with which most letters ended in that century, so some of the accusations and characterizations are certainly vastly exaggerated. Furthermore, Newton had curious ideas about scientific priorities. Today, publication is the dominant criterion, and independent discovery is a recognized, frequent occurrence. For Newton it was most important who first had an idea, even if kept secret, and he would give no credit to a later, independent discoverer.

Newton professed to abhor controversy, but the root of the matter was that he did not tolerate criticism. Before giving even a cursory account of the polemics in which Newton entered, it will serve well to read a letter by the philosopher John Locke a level-headed and generous man, a friend and admirer of Newton, to Lord King Oates on April 30, 1703. The letter is self-explanatory and shows to what length even his friends had to go when dealing with the great genius.

> Dear Cousin,
> I am puzzled in a little affair, and must beg your assistance for the clearing of it. Mr. Newton, in Autumn last, made us a visit here: I showed him my essay upon the Corinthians, with which he seemed very well pleased, but had not time to look it all over, but promised me if I would send it him, he would carefully peruse it, and send me his observations and opinion. I sent it him before Christmas, but hearing nothing from him, I, about a month or six weeks since, writ to him, as the enclosed tells you, with the remaining part of the story. When you have read it, and sealed it, I desire you to deliver it at your convenience. He lives in German St.: you must not go on a Wednesday, for that is his day for being at the Tower. The reason why I desire you to deliver it to him yourself, is, that I would fain discover the reason of his so long silence. I have several reasons to think him truly my friend, but he is a nice ["touchy"] man to

deal with, and a little too apt to raise in himself suspicions where there is no ground; therefore, when you talk to him of my papers, and of his opinion of them, pray do it with all the tenderness in the world, and discover, if you can, why he kept them so long, and was so silent. But this you must do without asking why he did so, or discovering in the least that you are desirous to know. You will do well to acquaint him, that you intend to see me at Whitsuntide, and shall be glad to bring a letter to me from him, or anything else he will please to send; this perhaps may quicken him, and make him despatch these papers if he has not done it already. It may a little let you into the freer discourse with him, if you let him know that when you have been here with me, you have seen me busy on them (and the Romans too, if he mentions them, for I told him I was upon them when he was here,) and have had a sight of some part of what I was doing.

Mr. Newton is really a very valuable man, not only for his wonderful skill in mathematics, but in divinity too, and his great knowledge in the Scriptures, wherein I know few his equals. And therefore pray manage the whole matter so as not only to preserve me in his good opinion, but to increase me in it; and be sure to press him to nothing, but what he is forward in himself to do. . . .

It is apparent that dealing with Newton was no simple problem.

The three most important people who had the bad luck to enter Newtonian controversies were Hooke, Flamsteed, and Leibnitz.

Robert Hooke was seven years older than Newton. Small and of frail constitution, he suffered various ailments all his life, but that did not prevent him from becoming a distinguished scientist in several fields. His largest published work, *Micrographia,* is a treatise on microscopy. He was versed in optics, mechanics, astronomy, and other fields. He was first curator and then secretary of the Royal Society, and one of his duties was to demonstrate weekly new experiments to the company. Hooke also had a difficult character and a propensity to claim ideas as his own, even on relatively weak grounds. As mentioned before, Hooke criticized Newton's optics paper in 1672, which was enough to bring about one of Newton's first outbursts; he was then only thirty years old.

Later, Hooke claimed, correctly, to have suggested the inverse-square law for attraction. Others had, too. However, he could not prove that this law could explain the elliptic planetary orbits, and this was the truly central issue. Newton was unwilling to give Hooke any credit. The diplomatic Halley, at the time of the publication of the *Principia,* tried to mediate between the two prima donnas. Newton then mentioned Hooke in the first edition of the *Principia,* but later, becoming more and more self-centered and self-righteous, he went to the length of seriously considering the suppression of the third book of the *Principia* for reasons set forth in the following letter to Halley:

Sir,
In order to let you know the case between Mr. Hooke and me, I gave you an account of what passed between us in our letters, so far as I could remember; for 'tis long since they were writ, and I do not know that I have seen them since. I am almost confident by circumstances, that Sir Chr. Wren knew the duplicate proportion when I gave him a visit; and then Mr. Hooke, (by his book *Cometa* written afterwards,) will prove the last of us three that knew it. I intended in this letter to let you understand the case

fully; but it being a frivolous business, I shall content myself to give you the heads of it in short, viz. that I never extended the duplicate proportion lower than to the superfices of the earth, and before a certain demonstration I found the last year, have suspected it did not reach accurately enough down so low; and therefore in the doctrine of projectiles never used it nor considered the motions of the heavens; and consequently Mr. Hooke could not from my letters, which were about projectiles and the regions descending hence to the centre, conclude me ignorant of the theory of the heavens. That what he told me of the duplicate proportion was erroneous, namely, that it reached down from hence to the centre of the earth. That it is not candid to require me now to confess myself, in print, then ignorant of the duplicate proportion in the heavens; for no other reason, but because he had told it me in the case of projectiles, and so upon mistaken grounds accused me of that ignorance. That in my answer to his first letter I refused his correspondence, told him I had laid philosophy aside, sent him only the experiment of projectiles, (rather shortly hinted than carefully described,) in compliment to sweeten my answer, expected to hear no further from him; could scarce persuade myself to answer his second letter; did not answer his third, was upon other things; thought no further of philosophical matters than his letters put me upon it, and therefore may be allowed not to have had my thoughts of that kind about me so well at that time.

Newton then goes on to say that he had revealed the inverse-square law many years before, in letters to Oldenburg and Huygens. Hooke had wrongly supposed that law to hold good down to the centers of the sun and planets. His own theorem on what happens is something that sets that matter straight and is something, he says, that Hooke could not possibly have guessed. "There is so strong an objection against the accurateness of this proportion, that without my demonstrations, to which Mr. Hooke is yet a stranger, it cannot be believed by a judicious philosopher to be anywhere accurate." He then informs Halley that he had designed the whole work to consist of three books, of which the second is ready, but,

> The third I now design to suppress. Philosophy is such an impertinently litigious Lady, that a man had as good be engaged in lawsuits, as have to do with her. I found it so formerly, and now I am no sooner come near her again, but she gives me warning. The two first books, without the third, will not so well bear the title of *Philosophiae Naturalis Principia Mathematica;* and therefore I had altered it to this, *De Motu Corporum libri duo.* But, upon second thoughts, I retain the former title. 'Twill help the sale of the book, which I ought not to diminish now 'tis yours. . . .

The relation between Hooke and Newton remained so strained that Newton vowed that he would not submit papers to the Royal Society as long as Hooke was its secretary. In 1703, when Hooke died, Newton became president of the society.

The quarrels with Flamsteed are another sorry chapter in Newton's life. Flamsteed was the son of a merchant, but by no means rich. He was also frail in body and was often forced to suspend his work for health reasons. He studied at Cambridge, where he met Newton and Barrow. He took holy orders in 1675 and devoted himself to a life of astronomical observations. He came to the attention of Oldenburg, and when the position of Astronomer Royal was established, following a proposal by the Royal Society, Flamsteed was appointed to it. The salary was £100

Isaac Newton in a study by the painter John Vanderbank for the portrait that he made in 1727, when Newton was eighty-five years old. (Royal Society, London.)

per year, but it was not paid regularly. Flamsteed was entrusted with the building of an observatory for his use at Greenwich, which he proceeded to do, using in part his own personal funds. He also built instruments and ground lenses with little help from others. Only in 1688, using a small inheritance from his father, could he afford a regular helper.

Flamsteed had a difficult character, and one of the objects of his suspicions was Halley. He saw in almost any of his actions some recondite motive against himself. Flamsteed's observational work was the best of his age and he took infinite pains about it. He did not mince criticism on the work of other contemporary astronomers whose observations were not as accurate as his, and, undoubtedly, he was not a diplomat, neither with Newton or with his colleagues.

Newton used Flamsteed's observations and relied on them in developing the lunar theory. There are innumerable letters from Newton requesting observational data and spurring Flamsteed to deliver them. Flamsteed possibly was a perfectionist; he was often sick and clearly did his best to oblige Newton. Newton treated him not as a colleague, but as a subordinate helper, offered him money to pay for an assistant, and offended Flamsteed many times. In the end, he obtained the observations he wanted and used them, quoting Flamsteed in the first edition of the *Principia*. Later he quarreled with the astronomer even more seriously and took advantage of his immense prestige and the official position he held to humiliate the worthy astronomer. He also forced the publication of Flamsteed's observations at a stage when the author did not consider them ready for publication. In the second edition of the *Principia* (1713), Newton omitted crediting Flamsteed, as he should have done. This pettiness and vindictiveness from such a great man as Newton was a sorry trait in his character.

The worst controversy of all was with Leibnitz over the discovery of the calculus. That dispute was preceded by a long period of goodwill and fair recognition of the reciprocal merits. The story is very complicated and much has been written on the subject. To follow all the details of the controversy would be useless and

Gottfried Wilhelm von Leibnitz (1646–1716), philosopher, mathematician, and a great figure of his century. He was one of the founders of the infinitesimal calculus and one of the earliest thinkers on modern mathematical logic and calculating machines. (Portrait by S. Scheits, Herzog Anton Ulrich Museum, Braunschweig.) Leibnitz's calculating machine in the Niedersächsische Landesbibliothek, Hannover. (Deutsches Museum, Munich.)

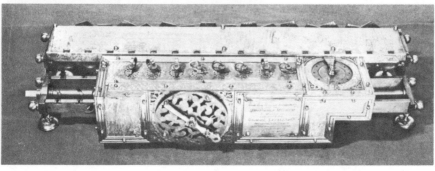

confusing. The war started around 1699, but referred to a large extent to events that had occurred about thirty years earlier. Unfortunately, national pride entered the controversy, and lesser persons on both sides fomented the fire and carried the offensive. Newton composed vicious tracts and attributed them, with their consent, to others. Furthermore, Newton, by then president of the Royal Society, appointed a committee that should have arbitrated the fight, as proposed by Leibnitz. However, the committee was stacked and acted in an extremely partisan way. The verdict of history is that the calculus was independently invented by both parties.

In his later years in London, Newton lived in state. He had been ennobled by Queen Anne in 1705, and he was now Sir Isaac; he was wealthy and universally considered the greatest living scientist. He presided over the Royal Society and in the twenty-four years of his tenure, he ruled it with an iron fist. Nobody could be elected without his consent. Under those conditions even such a great intellect succumbed occasionally to adulation, and he was ill served by his admirers in the polemics he carried on. He died on March 20, 1727, and is buried in Westminster Abbey together with other great Englishmen.

Chapter 3

What Is Light?

Newton's Heirs: Mathematics and Nature

Newton's legacy is of two types, symbolically represented by his major books — the *Principia* and *Opticks*. The *Principia* is a finished work and a paradigm for the way in which mathematical physics should be carried on. It is, in the main, a deductive work in which from a few laws — the laws of motion and the law of attraction — one derives numerous and grandiose consequences. The fabric of the universe and planetary motion are the most spectacular, but in Book II of the *Principia,* there are many applications to fluid dynamics, vibrations, and other subjects. They are the harbingers of things to come.

Opticks is an entirely different work, in which experiment is much more important than in the *Principia*. The conclusions are far from firm, and consistency is not the strongest feature of the work. Even if it is formally written in a deductive frame, imitating Euclid, the garb is only superficial. The final pages of the book, with its famous queries, is the place where the great man takes us into his confidence, relinquishing his superhuman aloofness.

The followers of Newton may be divided into those who developed the fabric of the *Principia* and those who explored new fields, chiefly heat and electricity. The first were more theoretically and mathematically inclined, the second were more experimentalist.

The Mathematicians

Starting with the first group, in Newton's time the invention of infinitesimal calculus provided a new and unsurpassed tool for attacking physical problems. There the advantages of the Leibnitz formulation and notation, combined with the stubborn veneration in England for Newton's work, gave the continental scientists an edge that was to last for about a century. For many decades there was no distinction between mathematicians and what we would now call theoretical physicists. The Bernoullis, a family that in four generations, from the early seventeenth to the late eighteenth centuries, produced eleven important mathematicians (four named Nikolaus, three Johann, two Jakob, and two Daniel) who often fought among themselves; the incomparable Leonhard Euler (1707–1783); and the great Alexis Claude Clairaut (1713–1765) and Jean le Rond d'Alembert (1717–1783) are all more important as mathematicians than physicists, but each of them also has some considerable physics result to his credit.

Hydrodynamics, the theory of elasticity, the beginning of the kinetic theory of gases, the study of waves and string vibrations, refined geodesy, and the development of Newtonian mechanics make up a partial list of accomplishments in which the new mathematics was brought to fruition. This grandiose use of mathematical techniques also set the example for future generations, and we still study Bernoulli's theorem in hydrodynamics, the Euler equations for a gyroscope, and d'Alembert's operator for waves.

Physics furnished to the mathematicians problems on which to test their mathematical discoveries or problems that stimulated their inventiveness. The mathematicians created methods that became as much a part of the physicist's equipment as many physical instruments, such as lenses and balances. Their importance was thus great, and we must keep their work in mind if we want to understand the evolution of physics. However, the mathematicians did not contribute much to fundamental discoveries of new phenomena, and, specifically, to the accumulation of facts concerning electricity, magnetism, and heat.

The protagonists moved in a background of eighteenth-century enlightment. Some were superior human beings, who one would have liked to have met, while some were spiteful and mean, and the greatness of their intellects contrasted with the pettiness of their characters. They often traveled between the courts of Frederick the Great of Prussia, Catherine the Great of Russia, and similar rulers. Their incomes came from princely generosity in the form of pensions or salaries as academicians. They usually published their works in the proceedings of academies, which reached their maximum splendor in that period, when they were the true repositories of science and the places in which new science was created. The universities in general were less important and mostly limited to a teaching function. In time the situation changed and the academies became more like honorary and publishing societies, while research shifted to the universities. The great academicians of the eighteenth century were also, by necessity, men of the world; many had outstanding literary or musical talents, and, at least in Berlin and St. Petersburg, they had to remain persona grata at court or change protectors. The princes, in turn, were inclined to spend money on the academies for reasons of prestige and hoped-for public utility. It was a time when each prince had to have his own little Versailles, his manufacturer of porcelain, his theaters, and his academy, much like today new members of the United Nations create their own airline and develop nuclear reactors. Some ministers, such as France's finance minister, Jean Baptiste Colbert, had hoped and expected economic advantages from an improved technology fostered by a better science. Some princes, for example, Frederick of Prussia, could even comprehend the works of their academicians.

For many of the academicians, mathematics was the main activity. Their results in pure mathematics were colossal, and they are still a large portion of what we study today. Euler's formulas for the trigonometric functions, his relation in topology, Jakob Bernoulli's (1654–1705) numbers and his distribution in probability, d'Alembert's theorem in algebra, and Clairaut's differential equation are only a small sample of their legacy. Any history of mathematics must devote chapters to them.

The effect of mathematics on physics was, however, unexpected. It turned out that theories created for purely mathematical reasons were applicable to physics. That surprising result is still a puzzle, although we may surmise some partial answers. For one, the mathematical theories invented by pure mathematicians without any consideration for their possible applications, give schemes of reasoning and models of inference that are suited to the human mind; the scientist then may find the phenomena that fit the paradigm. This does not, however, solve the puzzle of why the fit is so good. Eugene Wigner (1902–   ), the distinguished mathematical physicist who did so much to introduce group theory in modern physics, has written an essay, "The Unreasonable Effectiveness of Mathematics in the Natural Sciences," in which the reader will find interesting ideas. There is also the possibility that the fit is limited to certain parts of science, which for that very reason have grown to prominence. The very great domain of sciences, doctrines, or problems not accessible to mathematics remains in a shadow. Whatever the reason, the application of theories created for purely mathematical reasons, and without any idea of their applicability to physics, is a striking phenomenon that has continued for more than a century.

In modern times, group theory, tensor analysis, Lie groups, noncommutative algebras, matrix theory, and fiber bundles are only some of the chapters of pure mathematics first developed as such and later applied to physics. The physicist finds himself in need of progressively more abstract and sophisticated mathematics, as well as more refined observational instruments.

Examples of eminent mathematicians who have contributed to physics by the invention of mathematical formalisms include Joseph Louis Lagrange (1736–1813), Jean Baptiste Joseph Fourier (1768–1830), and William Rowan Hamilton (1805–1865). The first was born nine years after Newton's death, the last died when electricity was already in practical use. Their methods are still the stock in trade of current mathematical physics.

Lagrange, one of the greatest mathematicians that ever lived, recast Newtonian mechanics in his *Mécanique analytique* (1788). He devised general methods for solving all problems of Newtonian mechanics. No matter how complex the system studied, he gave uniform methods to translate the mechanical problem into differential equations. His methods apply to systems with a finite number of degrees of freedom, as well as to continuous systems, and there are rules to make the transition from one type of problem to the other. Lagrange was proud of the

Mathematician Joseph Louis Lagrange (1736–1813). In his Mécanique analytique, *he recast Newtonian mechanics in a form so general and abstract that it proved suitable for all further developments of mechanics — including relativistic and quantum mechanics. (Comando Scuola di Applicazione, Torina.)*

abstractness and generality of his methods; he bragged that his book did not contain any diagrams, because everything was entrusted to algebra. For many years the Lagrangean methods were accessible only to a few skilled mathematicians and did not have much impact on physics, but toward the middle of the nineteenth century, the methods entered the field of electricity through the work of Lord Kelvin, James Clerk Maxwell, and others. In modern times, their importance has progressively increased, and modern field theory always starts by postulating a "Lagrangean function." Furthermore, analytical mechanics turned out to be the formulation of mechanics best adapted to its transformation into quantum mechanics.

Lagrange's work was continued and expanded by the Irishman Hamilton, who discovered a profound analogy between the paths of rays of light and trajectories in mechanics. The same mathematical pattern and the equations connected with it describe both phenomena. That work, published around 1832, gave a general mathematical method for treating many phenomena. Its roots are in the minimization of certain functions of the coordinates and momenta. The power of these methods was revealed even more when they could easily be applied to relativistic mechanics and to quantum mechanics. In fact, they are now a common heritage and the terms *Hamiltonian* and *conjugate coordinates and momenta* are as important as *Lagrangean*.

In 1822 Fourier published *Analytical Theory of Heat,* a book summarizing his previous memoirs on the subject. The work deals primarily with heat conduction and considers heat as an indestructible fluid. Thus, it is irrelevant to thermodynamics. On the other hand, heat propagation forms the first example of contact action, as opposed to the Newtonian action at a distance. Fourier developed the mathematical tools for dealing with such continuous mediated actions: partial differential equation. It turned out that these equations were in part the same as those involved in potential theory, the mainstay of action at a distance. That indicated deep interrelations between the two approaches. Kelvin and Maxwell were influenced by this mathematical equivalence. Furthermore, Fourier's work also contained the first examples of a mathematical tool, orthogonal functions, which has become fundamental to quantum mechanics.

The Physicists

We come now to Newton's successors in the spirit of his *Opticks*. Although they obtained capital results, they are not as famous as the mathematicians. A large part of their theories proved ephemeral, and the important facts they discovered have been incorporated in later work, the original discoverers somewhat forgotten.

To make some order in this narrative, I will consider in succession optics, electricity, and heat. At the time of the death of Newton in 1727, optics was relatively more advanced than electricity or heat.

Joseph Fourier (1768–1830) was one of the French mathematicians who flourished at the time of the Revolution. He became a protegé of Napoleon, whom he followed to Egypt, and was a prefect under the Empire. His methods of analysis, including the use of Fourier's series and integral, have become central to much mathematical physics.

Newton had given us a tentative theory of light, and supported by experiments and by his immense authority, it dominated the thinking of physicists for almost a hundred years. When Newton formulated his optics, however, he had not gone unchallenged. Robert Hooke (1635–1702) had objections, and Christiaan Huygens (1629–1695) had a radically different conception of the light phenomena. The Dutch physicist was convinced that they had something to do with waves, and by intuition he had grasped a central point when he formulated the Huygens principle. He had understood the subtle and difficult mechanism that reconciles wave propagation with rectilinear rays, but had missed the much simpler concept of periodicity and interference, which form the basis for his own great discovery. The study of double refraction had been explained by him on the basis of his principle and strengthened his point of view. Nevertheless, Newton's authority, and the fact that he had explanations for colors and other phenomena for which Huygens had none, carried the day, and, with a few exceptions, notably that of Euler, the corpuscular theory was accepted.

Electricity, which is the kingpin of classical physics, was in its infancy. The properties of magnets, the attraction they exercise on iron, the use of the compass as an aid to navigation, and the attraction exercised by amber rubbed with silk or fur on light bodies were all facts known for centuries. Newton was well aware of electric phenomena, and he mentions them specifically in the *Principia* and in the queries of *Opticks*. However, anybody who has experimented in electrostatics knows how fickle it is. Humidity affects the phenomena and, at first, they are hardly reproducible. It takes some doing before they can be demonstrated without a high risk of failure, even when one knows what one is looking for. How galling and confusing experiments in electrostatics must have been in the eighteenth century! And still experiment was absolutely necessary, even for posing sensible problems to be solved.

Heat was in an intermediate situation. Thermometry was in hand, and some of the fundamental properties of gases were known. The concepts of specific heat and latent heat were on the horizon.

Light Is Waves: Thomas Young, A Universal Talent

We have to reach the time of Napoleon to see a radical progress in experiments and ideas concerning optics. First, in England Thomas Young (1773–1829) clearly understood and demonstrated the interference principle, forcing a wave interpretation of luminous phenomena. In a paper published in the *Philosophical Transactions of the Royal Society* in 1802, entitled "An Account of Some Cases of the Production of Colours, not hitherto described," he says,

> The law is that wherever two portions of the same light arrive at the eye by different routes, either exactly or very nearly in the same direction, the light becomes most intense when the difference of the routes is a multiple of certain length, and least intense in the intermediate state of the interfering portions; and this length is different for light of different colours.

Thomas Young (1773–1829) was a universal genius who left results of permanent value in fields as diverse as physiological and physical optics and Egyptology. (Courtesys of the University of California, Berkeley.)

As a model, Young had in mind sound waves, as well as waves in a liquid surface, as he had demonstrated at the Royal Institution. He performed a decisive experiment obtaining interference of light waves coming from two pinholes at a small distance. He could thus measure the wavelength of light, finding 0.7 micrometers for red and 0.4 for violet.

That great achievement assured Young's fame. He also considered diffraction phenomena and the color of thin plates from the point of view of a wave theory, and he brought qualitative clarity everywhere, but not a refined quantitative verification.

The work was appreciated by his contemporaries, and in 1803 Young gave the Bakerian Lecture of the Royal Society on "The Theory of Light and Colour." This lectureship is one of the great honors of British science and, in later times, we find among the lecturers Humphrey Davy, Michael Faraday, Lord Kelvin, James Clerk Maxwell, and Ernest Rutherford.

Young was born at Milverton, Somerset, England, into a well-to-do family of Quakers, by profession bankers and textile merchants. He was the first of ten children, a prodigy who learned to read at age two. By the age of fourteen he had written an autobiography in Latin, one of the many languages he knew by then. At school he read, in the original, the Latin and Greek classics, as well as Italian and French writers, making copious notes on what he read always in the original language of the authors. He further extended his studies to oriental languages—Hebrew, Persian, Arabic, and others. He also studied Newton's *Principia* (quite an enterprise by itself), Lavoisier's *Elements of Chemistry,* and other scientific works.

His choice of a profession was influenced by an uncle who was a prominent physician, and by his family's approval to turn to a medical career. Thus, at the age of nineteen, he moved to London to study medicine. In London he had access to high-placed persons. He frequently saw the statesman Edmund Burke, Sir Joshua Reynolds, the painter, and several members of the aristocracy, and he started to become a man of the world, loosening his early ties to Quakerism. He studied the mechanism of accommodation of the eye, and in 1794, at twenty-one, he was elected a Fellow of the Royal Society! From London he went to Edinburgh and Göttingen, where he pursued medical studies. He returned to England in 1797 and went to Emmanuel College at Cambridge to further study medicine. At that time his comrades at the college called him "Phenomenon Young," with a mixture of derision and respect. Shortly after his admission to Emmanuel College, Young was in London visiting his physician uncle when the latter died, leaving Young a large estate, including houses, books, art, and £10,000 in cash. The large inheritance made him financially independent and helped support him for the rest of his life. Young finished his studies at Cambridge in 1799. By then he had read some of the major mathematicians: Euler, the Bernoullis, d'Alembert, and noted their work on vibrating cords. He advanced some ideas in the course of that study, only to find that he had been anticipated by many years by the continental mathematicians.

Young started practicing medicine in London in 1799. Medicine at the time was nearly ineffectual and diagnostic means were poor. Public health left much to be desired, and it was in that field that it would have been possible to make rapid progress, as shown by Edward Jenner's discovery of smallpox inoculation (1796). Young's professional standing was good, but not outstanding. He was probably too much of a scientist and not enough of a fashionable healer to become a leader in the profession. On the other hand, he wrote on the physiology and anatomy of the eye, obtaining results of permanent value on color vision. By 1803 Young was well known as a physicist, and had been appointed Professor of Natural Philosophy at the Royal Institution in London. The Royal Institution is a unique institution created by the initiative and money of the American-born Benjamin Thompson, later Count Rumford (1753–1814). (The Institution was Faraday's home for all his career, and I will describe it in Chapter 4.) Young's stay at the Royal Institution was brief, merely three years. He did not have the lecturing gifts required for the job and often spoke above the audience's level of understanding. Later, he compiled an important book in two volumes, *A Course of Lectures on Natural Philosophy and the Mechanical Arts* (1807), which was based on his lectures. On leaving the Royal Institution, Young devoted more time to medical practice. In 1802 he had been elected foreign secretary of the Royal Society, and he retained that office for life.

Publications by Young cover an incredible variety of subjects — physiological optics, theory of the rainbow, fluid dynamics, capillarity, naval architecture, gravity measurements with the pendulum, tides theory — an incomplete list of just the physics topics. Young later collaborated with the *Encyclopaedia Britannica* and offered to write articles on the entries: alphabet, annuities, attraction, capillary action, cohesion, color, dew, Egypt, eye, focus, friction, halo, hieroglyphics, hydraulics, motion, resistance, ships, sound, strength, tides, waves, and anything medical — all subjects on which he had written original papers.

By 1814 Young discovered another interest—hieroglyphics. The famous bilingual Rosetta Stone had been discovered in 1799 during Napoleon's Egyptian expedition. When Napoleon had to evacuate Egypt, the stone was brought to London. Young became acquainted with it in 1814. Although others had studied it before him, he succeeded in making a capital advance; he found that certain words were written phonetically. That was a key to the interpretation, which was done in part by Young and, more completely, by Jean Champollion, a French professional Egyptologist. As in the case of optics, Young opened a subject that was elaborated on and brought to fruition by others.

In later years Young was an accomplished man of the world and his politeness was not only formal. In his relations with the great followers of his ideas—Augustin Jean Fresnel (1788–1827) and Champollion—he was, on the whole, fair and even generous, except for some occasional manifestation of a patronizing attitude. With Champollion he exchanged polemics, due more to Champollion than to him. With Fresnel, the relations were cordial.

Young went on to introduce another fundamental idea on light. When the lack of interference between two rays refracted by a calcite crystal had everybody in a quandary, Young came out with the suggestion that optical waves were transversal and that the polarization was connected with the direction of motion of the vibrations which is perpendicular to the direction of propagation. Lights polarized in perpendicular planes cannot extinguish each other. It was a simple idea, and it occurred independently to Fresnel too, but later.

From about 1815 on, Young was submerged in official occupations. The English government had to do something about the sorry state of their system of units of length and mass. It could hardly follow the admirable work done by the French during the Revolution, because national pride and narrow self-interest prevented it. Thus, they tried to define the inch by connecting it to the length of a pendulum striking one second. Young participated in this work. The conclusions were less than admirable, but they were adopted by an act of Parliament in 1824.

Next, he worked on life insurance problems. He had become an inspector of calculations, that is, a chief actuary, for an important insurance company at a handsome salary. In 1818 he was appointed superintendent of the *Nautical Almanac,* and undertook work to improve practical astronomy and navigational help. His extremely active life ended at age fifty-six, in 1829.

Young, the great amateur, was an extraordinary universalist of powerful imagination, but perhaps limited staying power; his rival in the theory of light, Fresnel, was a very different type—a great perfectionist who, once launched on a subject, did not leave it until it was completely finished. Fresnel became the supreme optician of the nineteenth century.

The French Scientific Nurseries

France and England formed their scientists in different ways. In England, scientific education was imparted at the universities—mainly Cambridge and Oxford—and through the method of apprenticeship for engineering, medicine, and professional disciplines. University ties to the Church were strong. Protestant divines could marry, and not a few scientists took holy orders or were related to clerical families. There was also a tradition in the aristocracy of rich amateurs devoted to natural science. In general, the government was aloof from scientific endeavors, and private financing was paramount, except in the fields of geography and exploration. The great Industrial Revolution, then in full swing, was not too much influenced by science. Practical people, such as James Watt and other great engineers, were its technical leaders, and financial considerations were at the forefront.

The social standing of scientists was fairly high. Newton had been ennobled, and the precedent was later followed in a few cases, for example, Sir Humphry Davy.

The Royal Society was an important factor in British scientific life. Sir Joseph Banks (1743–1820) was its president from 1778 until his death and ruled it with an iron hand, as Newton had done previously. Banks was a botanist and agriculturist who had traveled extensively on exploratory expeditions, including a circumnavigation of the earth with Captain James Cook, but his true importance was that of an organizer and promoter.

In France, before the Revolution, science was an occupation restricted to a small number of people of the upper social classes. It was hard to climb socially and economically on the basis of scientific merit, and education was limited to the upper classes and the clergy, although exceptional talent could emerge, as shown by the humble origins of several academicians. The Jesuits had made great strides in creating a modern system of education. Indeed, some features of their schools are recognizable in today's eductional systems. Furthermore, they produced many excellent scientific texts.

The Revolution changed this situation. It is true that in its excesses it killed indiscriminately, and the death of Lavoisier and the Marquis de Condorcet, among others, are eternal reminders of its cruel and unthinking demagoguery. On the other hand, it radically improved the educational system of the country. Of course, there had been technical schools under the ancien régime, but they were addressed (outside of the clergy and the law) to the education of military and civil engineers and doctors. Admission and advancement were dominated by criteria of nobility, ability, and connections, which functioned in unforeseeable fashion. After the execution of the King, France, under assault by the rest of Europe, was in dire need of technical help, and the Convention founded new schools, the most important of which were the École Normale and the even more selective École Polytechnique. The second became the cradle of scientific and technical knowledge in France. Admission was based on stiff competitive examinations, strict meritocracy prevailed, and the instruction was very strong in mathematics, which slanted toward practical application. Military discipline was enforced on the unruly, intelligent, highly motivated pupils. One of the chief organizers of the school was Gaspard Monge (1746–1818), a commoner, who had gone through the military schools of the ancien régime on his sheer ability, which attracted the protection of his superiors.

Napoleon visits the École Polytechnique in 1815, during the Hundred Days Empire, trying to rekindle the students' enthusiasm. (From École Polytechnique, Livre du centenaire, *Paris, 1895.)*

Monge, a distinguished geometer, became very friendly with Napoleon. The École Polytechnique, republican in origin, had to transform its politics several times, but it managed to remain relatively unscathed. According to Dominique François Jean Arago (1786–1853), a well-informed contemporary, when Napoleon transformed the Republic into an Empire, he found manifest disapproval from many of the students. Incensed, he planned punitive measures against the students and the school. Napoleon said to Monge, "Well, Monge, almost all your students are revolting against me!" To which Monge replied, "Sir, we have had trouble enough into making them into good republicans; give them time to become favorable to the Empire. Moreover, allow me to say it, you have also turned rather suddenly." Monge could talk that way to Napoleon and protect his pupils because he was an old trusted friend. Napoleon, however, early in the Empire, conflicted with and transformed the École Polytechnique. Napoleon had respect for science, or at least he appreciated its practical importance. When he made his expedition to Egypt, he took with him a host of scientists, among them Monge. He flaunted his membership in the National Institute, a reincarnation of the French Academy, which had been suppressed during the Reign of Terror, by signing his proclamations "Bonaparte, Member of the Institute and General in Chief."

At the turn of the century Lagrange; Pierre-Simon Laplace (1749–1827); Adrien-Marie Legendre (1752–1833), who had been aloof from the Revolution; as well as former revolutionaries like Lazare Carnot (1753–1823), Monge, and Fourier were major figures in Paris. They ruled on scientific matters and taught or were connected with the École Polytechnique. Members of the younger generation, such as Fresnel, André-Marie Ampère (1775–1836), and Sadi Carnot (1796–1832) could turn to them, with some awe. It was on the whole an extremely lively society, where military adventure, travel to faraway countries, and high scientific achievement mixed to make life exciting. Most of the time France was at war with England, but, apparently, war did not prevent travel for scientists, and we find the distinguished Englishmen Davy, visiting Paris and being honored by his French colleagues.

A reflection of that life emerges through the biographies of the deceased members of the Académie des Sciences written by Arago, its secretary. It is possible that Arago may have slightly embellished the facts, since, reading his autobiography, Alexandre Dumas and the Three Musketeers kept coming to my mind.

One important achievement of the French Revolution that had repercussions for all industrialized nations was the introduction of the metric system. It was planned by a commission of top scientists, including Lavoisier, Lagrange, Monge, Condorcet, and Laplace. Napoleon's armies propagated the system to the rest of Europe, although final adoption occurred only many years later. Only England and the English-speaking countries stuck to the obsolete units, an act, perhaps, of patriotism, but certainly of very expensive consequences. The decimal system of numbers was obviously a great boon; the rational relation between units of length, surface, and volume was a great simplification. The choice of units, on the other hand, was arbitrary. It was desirable to have reproducible units connected to unchangeable standards. It was thought at first that the length of a pendulum having the period of one second at a suitable location, might be a good choice. Better measurements of the arc of meridian of the earth tilted the choice toward a connection to the dimensions of the earth. All this has now been superseded; an international agreement has been reached wherein the standard of length is the wavelength of a krypton spectral line, the standard of time is the period of an atomic transition in the cesium atom, and only the standard of mass is a physical object — a platinum weight preserved in Paris. Possibly in the future all units will be defined on the basis of universal constants, such as h and c, with practical realizations represented by atomic properties.

Fresnel's Perfection

The seeds planted by Huygens and Young blossomed in the optical works of Fresnel. Augustin Jean Fresnel was born at Broglie in Normandy in 1788 when the French Revolution was about to begin. His father was an architect, and his mother was a member of the Mérimée family, distinguished for her brother, Léonor, a noted

Augustin Jean Fresnel (1788–1827) in a portrait that his friends claimed was a close likeness. (Courtesy of the University of California, Berkeley.)

painter, and his son, Prosper Mérimée, a cousin of Fresnel and a well-known literary figure best remembered for his short novel *Carmen,* the subject of the famous opera.

Contrary to Young, Fresnel was late in his development and a poor linguist. However, at age nine he had already shown uncommon technical ability in developing, along scientific lines, peashooters and bows and arrows. His health was not strong, but at sixteen he could enter the École Polytechnique, and from there he went to the engineering school, the École des Ponts et Chaussées. He obtained a government job as a civil engineer, building roads and bridges in the French provinces. There, in complete isolation from the scientific world, he started to study the nature of light as an avocation. In 1814 he wrote to his brother (also Léonor), to whom he was very close, asking for books from which he could learn about the polarization of light. He did not suspect that ultimately he would have to write what he wanted to read.

In 1815 Napoleon returned from Elba, where he had been confined by the European powers after his defeats of the previous year. A tremendous wave of enthusiasm shook France, countered by equally vehement feelings from Napoleon's adversaries. Fresnel was one of them and enlisted against Napoleon. The reestablished Hundred Days Empire fired him from his job and confined him first to Nyons and later to the village of Mathieu. With the battle of Waterloo and the second return of the Bourbons, Fresnel was restored to active service toward the end of 1815.

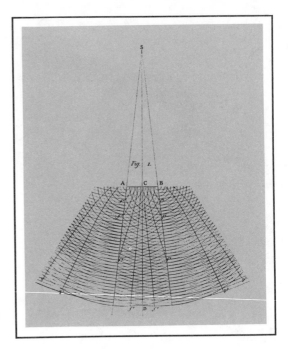

Fresnel's original drawing showing the interference patterns of light coming from two point apertures near each other. For the experiment to succeed, it is essential that the two sources be coherent. (Courtesy of the University of California, Berkeley.)

In the intervening months he had initiated the studies that were to revolutionize optics. He had observed the diffraction from a half plane and developed a careful theory for it, combining with mathematical skill the concepts of periodic vibration with a precise formulation of Huygens's principle. Fresnel was able to leave his enforced stay and visit Paris, where he went to see Arago, who rapidly appreciated him. Unfortunately, Arago had to tell Fresnel that his results, to a large extent, had been anticipated by Young. Fresnel's work, however, was more detailed and quantitative and had enough novelty to warrant publication in the *Mémoires* of the Académie des Sciences. The paper was shortly followed by a second on the same subject. Arago and the well-known mathematician Louis Poinsot (1777–1859) of gyroscope fame had been appointed as referees for Fresnel's work. They obtained a leave of absence for Fresnel from his superiors so that he could work for a few months in Paris, using Arago's facilities. In his studies at Mathieu, Fresnel had used apparatus he built with the help of the village smith, but diffraction studies call for fine mechanical apparatus—micrometers, slits, etc.—and he could hardly build them unaided.

Fresnel then passed from the study of diffraction to that of colors of thin plates. Here, too, Young had preceded him. In 1816 Arago, together with Joseph Louis Gay-Lussac (1778–1850) visited Young at his home in Worthing. The following is an account of the visit by Arago:

In the year 1816 I visited England, in company with my learned friend Gay-Lussac. Fresnel had recently made his debut in the career of the sciences, in the most brilliant manner, by his Memoir on Diffraction. This work, which, in our opinion, contained a capital experiment irreconcilable with the Newtonian theory of light, became naturally the first subject of our conversation with Dr Young. We were astonished at the number of restrictions which he imposed upon our commendation of it, when at last he declared that the experiment which we valued so highly was to be found, since 1807, in his Lectures on Natural Philosophy. This assertion appeared to us unfounded, and a long and very minute discussion followed. Mrs Young was present at it, without offering to take any part in it—as the fear of being designated by the ridicule implied in the sobriquet of *bas bleus* makes English ladies reserved in the presence of strangers; our neglect of propriety never struck us until the moment when Mrs Young quitted the room somewhat precipitately. We were beginning to make our apologies to her husband when we saw her return with an enormous quarto under her arm. It was the first volume of the Treatise on Natural Philosophy. She placed it on the table, opened the book without saying a word, at page 387, and showed with her finger a figure where the Curvilinear Course of the diffracted bands, which were the subject of the discussion, is found to be established theoretically.

The Académie des Sciences in 1818 opened a competition for a paper devoted to a theoretical and experimental investigation of light diffraction. The judges were Laplace, Jean-Baptiste Biot (1774–1862), and Siméon-Denis Poisson (1781–1840), who were all partisans of the emission theory, and Arago and Gay-Lussac, who were inclined to the wave theory. There were two competitors, Fresnel being the important one. Poisson noted a curious consequence of Fresnel's theory—at the center of the shade projected by a disc, there should be a luminous point. The result seemed paradoxical, but when the experiment was performed, the bright point was there. Needless to say, Fresnel's paper received the prize, and Poisson was converted.

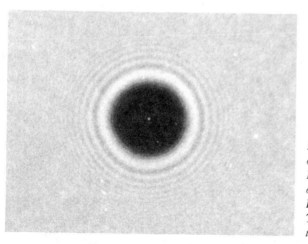

Diffraction of light by a disk, obtained by Fresnel's method. Note the bright spot at the center of the shadow, which so impressed Poisson. (Courtesy of Brian J. Thompson, University of Rochester.)

Another difficulty for the wave theory appeared in the interference of polarized light. Newton had asked in Query 26 of his *Opticks,* "Have not the rays of light several sides, endued with several original properties?" The phenomena of double refraction had prompted the query. For Young and Fresnel, the interference of polarized light presented problems until it dawned first on Young and later but independently on Fresnel that the vibrations of light differed in one respect from the vibrations of sound in air. In these the movements of the molecules are parallel to the direction of propagation of sound, so that the air is alternately compressed and rarefied, while light vibrations are transversal, as are the displacement of the points of vibrating strings. It is clear, then, that displacements in directions perpendicular to each other cannot interfere, and only displacements in parallel directions can.

Light can be polarized by double refraction: The two rays emerging from a calcite crystal are polarized perpendicular to each other. A new way of polarizing light was discovered by Étienne Louis Malus (1775–1812), another former pupil of the École Polytechnique who had accompanied Napoleon to Egypt. In 1808 he was observing from his house sunlight reflected by a windowpane in another building through a calcite crystal. To his surprise he saw only one image and not two as he had expected. He soon traced the effect to the polarization of reflected light — a new phenomenon. Malus was a firm believer in the corpuscular theory and interpreted his observations on that basis, although it required strange ad hoc hypotheses.

Fresnel combined all those observations in a complete theory of polarized light, including the concepts of coherence and elliptical polarization. He found the wave surface in a crystal and the laws governing the intensity of reflected and refracted light. All this was a major accomplishment that established the phenomenology to be explained. The crowning success would have been the observation of the properties of the medium that propagated light in vacuum — the ether. There Fresnel met insurmountable difficulties. This book is not the place to expand on all the paradoxes presented by the conception of the ether. Their hisory is tangled, but the conclusion is simple: The ether does not exist, or at least the properties of the vacuum are radically different from those of an elastic medium. Faraday and Maxwell were to make the first step toward unraveling the problem by showing that light is electromagnetic; Einstein would later demonstrate that the classical ether was unnecessary; quantum mechanics is now endowing the vacuum with strange and unexpected properties; and the end is not yet in sight. The mechanical theories are, however, dead.

The transversal nature of light vibrations appears to us almost trivial, but when the idea was formulated, it met great opposition. Even Arago, an ardent partisan of the wave theory, and a friend of Fresnel, refused to sign the paper in which Fresnel put forward that idea, the reason being that, in order to have transversal waves, the ether should have, according to any mechanical model, a rigidity that seems incompatible with all observations.

Fresnel died of tuberculosis at Ville d'Avray, near Paris, in 1827 at the age of thirty-nine. In the last years of his life, he was employed by the Lighthouse Commission of France. He made important practical inventions there, such as the

Fresnel lenses, which we can now buy in a plastic version in hobby stores. All his life he had been an extremely conscientious man, deeply religious in the Jansenist tradition. He was also rather shy. In a letter to his brother, he says, "I hardly find anything as painful as to have to drive people, and I confess that I do not know how to do it."

In 1823 he was elected to the Académie des Sciences, and the Royal Society of London selected him as a foreign member. The Académie reciprocated by electing Young as one of its eight foreign associates.

The scientific and personal relation of Young and Fresnel is interesting and important. It is well summarized in a letter from Fresnel to Young. The latter had invited him to write an article on light for the *Encyclopaedia Britannica*. This is in itself significant. Fresnel accepted first, but he later had to renounce the undertaking because of his grave illness. He then wrote a long letter to Young, in which he said, in part:

It seems to me, however (perhaps my amour propre blinds me), that what you left me to do in those parts of optics was as difficult as what you yourself had done. You had gathered the flowers, may I say with English modesty, and I have dug painfully to discover the roots.

I am far from laying claim to what belongs to you, Monsieur, as you have seen in the Supplement to the French translation of Thomson's Chemistry, as you will see also in the article I have just prepared for the European Review. I have declared with sufficiently good grace before the public, on several occasions, the priority of your discoveries, your observations and even your hypotheses. However, between ourselves, I am not persuaded of the justice of the remark in which you would compare yourself to a tree and me to the apple which the tree has produced; I am personally convinced that the apple would have appeared without the tree, for the first explanations which occurred to me of the phenomena of diffraction and of the coloured rings, of the laws of reflection and of refraction, I have drawn from my own resources, without having read either your work or that of Huygens. I noticed for myself also that the difference of path of the ordinary and extraordinary rays on emerging from a crystal plate was equal to that of the rays reflected at the first and second surfaces of a film of air which gives the same tint in the coloured rings. It was when I communicated this observation to M. Arago that he spoke to me for the first time of the note which you had published two years before on the same subject and to which, until then, he had not paid much attention. This does not, of course, entitle me to share with you Monsieur, the merit of these discoveries, which belong to you exclusively by priority: also I have thought it useless to inform the public of those things which I discovered independently but after you; and if I speak of them to you, it is only to justify my paradoxical proposition, *that the apple would have come without the tree*. For a long time, Monsieur, I have wanted to open my heart to you on this subject, and to show you exactly the extent of my claims.

Let us admit that my amour propre is too exacting and that I have received justice in your country (for I am perhaps one of the Frenchmen who have least to complain of at the hands of your countrymen), I am not the less astonished, I would almost say revolted, by what is so often reported to me of the shocking partiality with which your scientific journals exalt the most insignificant English discoveries over the most remarkable French ones. Certainly I should be the last to minimise your undeniable superiorities, especially in the realm of politics; but you will allow at least that we are far ahead in impartiality and love of justice.

This letter will perhaps appear to you, Monsieur, the outpouring of a sick man tormented by bile, and whose amour propre has been wounded by the neglect of his work in your country. I am far from denying the value which I would attach to the praise of English scientists and from pretending that such praise would not have been agreeable. But, for a long time this sensitiveness or this vanity which is called love of fame has been much blunted in me; I work far less to impress the public than to obtain that self approval which has always been the sweetest reward of my efforts. Without doubt I have often needed the spur of vanity to stimulate me to pursue my researches in moments of disgust or discouragement; but all the compliments which I have been able to receive from MM. Arago, Laplace, or Biot have never given me as much pleasure as the discovery of one theoretical truth and the confirmation of my calculations by experiment. The smallness of the effort I have made to secure the publication of my memoirs, of which only extracts have appeared, shows that I am not tormented by a thirst for fame, and that I have enough philosophy to prevent me from attaching too much importance to the enjoyment of vanity. But it is useless for me to spread myself further on this subject in writing to one who is himself too superior to be a stranger to this philosophy and who will readily understand me.

Young's reply to this and similar letters is unknown, but in published writings, he says:

I had first the pleasure of hearing at a meeting of the Academy of Sciences, an Optical paper read by Mr Fresnel, who, though he appears to have rediscovered, by his own efforts, the laws of the interference of light, and though he has applied them, by some refined calculations, to cases which I had almost despaired of being able to explain by them, has, on all occasions, and particularly in a very luminous statement of the theory, lately inserted in a translation of Thomson's Chemistry, acknowledged, with the most scrupulous justice, and the most liberal candour, the indisputable priority of my investigations.

I believe that Fresnel is too modest in his letter, but that the situation on the whole is fairly portrayed by the principals. That is an uncommon example that does honor to both.

Fresnel's work was practically limited to optics. In that he differed from many of his contemporaries who covered a great variety of subjects. In optics, however, he was superb. Few papers of his times can still be read as living physics, as his can. He is perfect in all details, and he reminds me of a jeweler rather than a sculptor. If Faraday may be compared to Michelangelo, Fresnel would be a Benvenuto Cellini. He was a perfectionist, and whatever he touched pretty much received its definitive treatment.

Fresnel's example influenced a number of important French physicists who continued his work in optics. Léon Foucault (1819–1868) and Armand-Hippolyte-Louis Fizeau (1819–1896) measured the velocity of light in air and water and found that it was greater in air (1850), a crucial experiment in favor of the wave theory of light and against the corpuscular theory. They also determined the absolute value of the velocity of light, a constant that is of immense importance in establishing the electromagnetic theory of light and later for relativity. The two physicists were friends for a while, and at times they collaborated, while at other times they worked

Joseph von Fraunhofer (1787– 1826), whose practical achievements and scientific acumen in the field of optics enabled him to rise from very humble origins to a position of nobility and prominence. A contemporary of Fresnel, he influenced the optical industry in Germany for at least a century. (Deutsches Museum, Munich.)

independently. Their optical work was continued by Marie Alfred Cornu (1841– 1902), and ultimately influenced A. A. Michelson (1852–1931), establishing a direct line from Fresnel to modern times, reinforced by Hendrik Lorentz's (1853– 1928) deep study of Fresnel.

The Messages of the Spectra: Fraunhofer, Bunsen, and Kirchhoff

In the nineteenth century, there was a figure contemporary to Fresnel who must be mentioned: Joseph Fraunhofer (1787–1826). The dates of his birth and death almost coincide with those of Fresnel, and he, too, died of tuberculosis, a scourge of nineteenth-century Europe, but how different were his family circumstances from those of Fresnel! He was born at Straubing, near Munich, the eleventh son of a glazier. His parents were very poor and uncultivated. Joseph became an orphan at age eleven and was apprenticed to a mirror-maker. When he was fourteen years old, the building in which he was working collapsed, covering him beneath the ruins, but his fortunes suddenly turned. The Elector of Bavaria, touched by the tragic event, gave him enough money to allow him to leave his apprenticeship and enter school. The Elector also recommended him to Joseph von Utzschneider, an important

Prince Maximilian of Bayern rescues the young glassworker Fraunhofer from a collapsing house. From a contemporary engraving. (Deutsches Museum, Munich.)

industrialist and politician who had an optical business among his many enterprises. Fraunhofer, at the time almost completely unschooled, showed a true passion for optics, and Utzschneider employed him in the optical business. Fraunhofer rose rapidly to be a partner. Utzschneider the entrepreneur had recognized the immense value of the technical gifts of his associate and helped him to secure a good financial position.

Fraunhofer soon realized the strict connection between the quality of the glass used and the performance of finished optical instruments. He then studied how to improve glassmaking in the firm's glass factory near Munich. The Swiss glassmaker Pierre Louis Guinand was probably the best glassmaker in Europe, and Utzschneider had hired him, but the glass produced was not as good as desired, and in 1811 Fraunhofer took over, with excellent results. Fraunhofer, convinced of the advantages of passing from empirical artisanship to refined scientific planning, studied the aberrations of lenses, chromatic and otherwise. In precisely measuring the refractive index of glass for different wavelengths, he rediscovered the black absorption lines present in the solar spectrum (William Hyde Wollaston, 1766–1828, had already seen some of them in 1802) and made a catalog of these "Fraunhofer lines," to be used as benchmarks. The figures of the solar spectrum etched personally by Fraunhofer are an admirable testimony to his multiple skills.

The prism spectrograph that Fraunhofer used to measure the refractive index of glass and (around 1815) the black absorption lines of the solar spectrum. (Deutsches Museum, Munich.)

The result of those efforts was that his optics became world famous and unsurpassed. The mechanical part of the instruments was to equal the optics, and Fraunhofer tackled the engineering problems. The Dorpat refractor, with a lens of 24 cm diameter and a weight of 1,000 kg, is one of his masterpieces. Fraunhofer's merits were recognized by the king of Bavaria, who ennobled him. The Optical Institute, the name of the Utzschneider and Fraunhofer's firm, hired more than fifty workers and became the leading optical firm in the world. The example set by Fraunhofer in combining theoretical optical work, glassmaking, and mechanical precision and prowess influenced the German optical industry in a lasting way. Either directly or indirectly, many of the famous opticians, such as Joseph Max Petzval (1807–1891), Karl August Steinheil (1801–1870), and Ernst Abbe (1840–1905), and optical firms, such as Zeiss and Leitz, descend from Fraunhofer.

Fraunhofer can also be considered one of the founders of spectroscopy, for the discovery of the absorption lines in the solar spectrum, for the recognition that they corresponded to emission lines in sparks and flames, and for the use of diffraction gratings, which he developed in various forms. The refinement of the art of spectroscopy occurred, however, mainly in the United States through Rowland and Michelson, fifty years after Fraunhofer's death.

The immense importance of spectral lines, however, escaped Fraunhofer and other physicists for many years, until about 1860, when their significance as signals

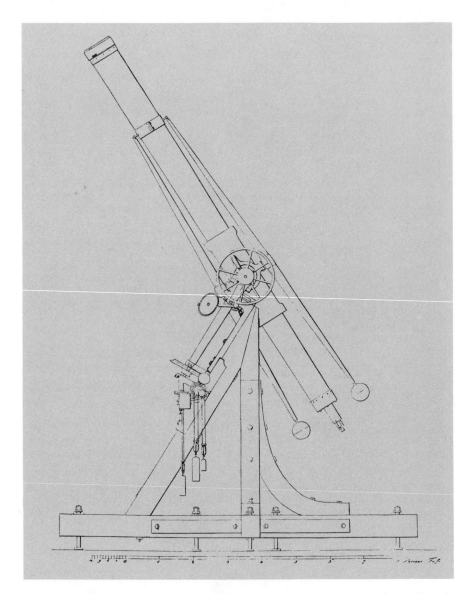

The refracting telescope Fraunhofer built for the Dorpat Observatorium in 1824. Focal length, 4.11 m; lens aperture, 24 cm. (Deutsches Museum, Munich.)

of atomic or molecular species emerged, mainly through the work of Robert Wilhelm Bunsen (1811–1899) and Gustav Robert Kirchhoff (1824–1887). Bunsen and Kirchhoff, both socially and in their careers, represent typical examples of German science at its best. Both came from families with a tradition of state

Robert Wilhelm von Bunsen (1811–1899) is famous for his practical inventions (such as the laboratory gas burner and his battery), as well as for his work in inorganic chemistry and photochemistry. His great achievement, which he shares with Kirchhoff, was the development of spectral analysis, which led to the discovery of cesium and rubidium. (Courtesy of AIP Niels Bohr Library, W. F. Meggers Collection.)

service and intellectual activity, and both devoted their lives to science in the congenial and prosperous atmosphere of the best German universities of the nineteenth century. Bunsen was thirteen years the senior of the two. A brilliant career as an inorganic chemist had brought him by 1852 to the University of Heidelberg. His interests were very wide and practical at the same time. The burner that carries his name is a permanent fixture of all chemistry laboratories, but, in times past, his type of battery was equally popular and important. Other instruments, such as the ice calorimeter, also bear witness to his experimental ingenuity. His accomplishments, however, extend well beyond the creation of new instruments. Photochemistry owes him some fundamental results; analytical chemistry, as we shall see, was transformed by him; and metallurgy, too, bears his imprint.

Gustav Robert Kirchhoff (1824–1887) is remembered for his many contributions to physics. The distribution law of currents in circuits, the radiation law, spectral gens's principle are among them. A major exponent of classical theoretical physics, he taught primarily at Heidelberg and Berlin and wrote several influential books. (Courtesy of AIP Niels Bohr Library, W. F. Meggers Collection.)

Kirchhoff studied at Königsberg, where there was a strong school of theoretical physics headed by F. E. Neumann (1798–1895). At the age of twenty, Kirchhoff solved the problem of calculating the distribution of currents in an arbitrary network, establishing what we call Kirchhoff laws. The simple but very important method and result is still one of the mainstays of practical electricity. He also applied his mathematical skills to the solution of problems of propagation of electricity, using Neumann's generalizations of Ampère's law to moving charges. Experimental results by Wilhelm Weber (1804–1891) and Rudolf Kohlrausch (1809–1858) determined a universal constant that appeared in the formulas and represented the velocity of propagation of electrical action. Kirchhoff noted that it practically coincided with the velocity of light, but the fundamental significance of that fact seems to have escaped him.

In 1850 Kirchhoff was appointed professor at Breslau, where he met Bunsen. They became friends, and when Bunsen transferred to Heidelberg, he managed to obtain an appointment for Kirchhoff at the same university. By 1859 Bunsen was using as an analytical tool the color imparted to the flame of his burner by salts. To improve the discrimination, he used colored glasses or colored solutions, through which he looked at the flame. His friend Kirchhoff suggested he use the spectroscope. Major discoveries followed. It became apparent that Fraunhofer's lines were characteristic of chemical elements and that the black absorption lines in the solar spectrum corresponded exactly in wavelength to bright emission lines obtained in emission by placing appropriate salts in the flame of a burner. The way was thus opened for spectral analysis, allowing us to detect the presence of certain elements in the sun, in the stars, and in other inaccessible sources. On the other hand, the relation between absorption and emission had to be clarified.

The power of spectral analysis was soon demonstrated by the discovery of new chemical elements. Bunsen used his superb analytical skills to obtain salts whose spectra would not show the ubiquitous yellow lines of sodium impurities. In the residues of some mineral water, he and Kirchhoff detected a new blue spectral line never observed before, and they attributed it to a new element, an alkali metal, they called cesium ("blueish" in Latin, from the color of the spectral lines). A few months later, Bunsen discovered another alkali metal, rubidium ("red," from the color of the characteristic spectral lines) in the mineral lepidolite. Spectral analysis became the most powerful analytical tool, and several chemical elements were discovered through its use — thallium, indium, and gallium were next.

The question of the relation between emission and absorption was tackled by Kirchhoff. He observed that the light emitted by a very hot body filtered through a flame containing sodium showed black absorption lines at the same wavelengths at which the flame by itself showed bright emission lines. To a modern physicist, that behavior suggests a resonance phenomenon, and the idea was put forward in Kirchhoff's time. Kirchhoff, however, wanted to be on safe ground, and atomic models or too detailed explanations did not afford enough reliability. On the other hand, thermodynamics was safe. He thus tried to apply this abstract but safe doctrine to the phenomenon and reached a very important result: Call emissive power, or emittance, e, the power emitted per unit surface in a unit frequency interval by any body, and call absorbing power a, the absorbed fraction of power incident on a body in a unit frequency interval. Kirchhoff showed that the ratio of e to a is a function of the frequency v and the temperature T, independent of the nature of the body. For a body perfectly absorbing all radiations, $a = 1$ and $e(v, T)$ is a universal function independent of the nature of the body (a so-called black body), provided it has $a = 1$. Clearly the determination of $e(v, T)$ was an important problem.

Bunsen and Kirchhoff had obtained the appointment of their friend Hermann von Helmholtz (1821–1894) at Heidelberg in 1858. When Helmholtz moved to Berlin in 1871 to become the head of the biggest physics institute of the Reich, he was nicknamed "the chancellor of German physics." Helmholtz invited Kirchhoff to Berlin as a professor of theoretical physics, but Bunsen, who had also been invited to Berlin, did not want to move from Heidelberg. In Berlin Kirchhoff taught a famous course in theoretical physics, which became a standard for many years. It had a tendency to formal perfection, an avoidance of models, and a love for rigor. It contrasted sharply with the English tradition. The lectures, however, do not seem to have inspired Heinrich Hertz (1857–1894) or Max Planck (1858–1947), the two most gifted students who attended them; apparently they were bored.

Kirchhoff's influence was certainly important in the choice of one of the central subjects of investigation of the Physikalisch-Technische Reichsanstalt, the great German institute directed by Helmholtz and inaugurated in 1888, one year after Kirchhoff's death. Ultimately, those studies on the black body led to the discovery of the quantum by Planck, in 1900. That is one of the major discoveries that signal the passage from classical physics to the new, twentieth-century physics.

Although Bunsen and Kirchhoff founded spectral analysis, no explanation connecting the spectra to atomic structure existed. Clearly, one had to devise an electromagnetic object that could emit the observed spectra. Maxwell's theory of electricity provided the tools for calculating the radiation by moving charges, but it was a far cry from that to a truly atomic or molecular theory of radiation. It ultimately turned out that classical physics was insufficient for the task, but it could point the way and lead far in the right direction. What could be accomplished within the boundaries of classical theory was explored by several authors, the most representative of which was Lorentz.

Chapter 4

Electricity: From Thunder
to Motors and Waves

The Difficult Conquest of Electrostatics
and Magnetism

Mechanics was the first part of physics developed in the pattern we use now. It served as a model for future work, and for a long time, there existed the desire and illusion of reducing all physics to mechanics. Electricity and magnetism form the other great pillar of classical physics, and, ultimately, they proved irreducible to mechanics. At the time of Newton's death in 1727, when mechanics had taken its modern shape, most of the facts of electricity were still to be discovered. The phenomenology of electrostatics and magnetostatics were unraveled to a large extent in the eighteenth century, and this chapter is devoted to it.

The ability of some bodies rubbed with cloth to attract light straws or bits of paper and some of the properties of the lodestone have been known since antiquity, but I shall not mention students of such phenomena before William Gilbert (1540–1603). Gilbert was an Englishman of well-to-do family. He studied medicine and mathematics at Cambridge, and later practiced medicine in London, eventually becoming so renowned that he was appointed physician to Queen Elizabeth I. He was thus a character of Elizabethan England; he might have seen Shakespeare performing at the Globe and worried about the Armada. The Queen regarded him highly and left him a special bequest for his studies.

Gilbert's work in magnetic phenomena is summarized in his great treatise *De magnete,* which he published in 1600. Modern study of electricity and magnetism starts with his experimental investigations, which lasted for more than fifteen years and cost him a good portion of his personal fortune. Before Gilbert, fantastic ideas prevailed, often bordering on magic. He performed many important experiments, such as building a "terrella," a little earth of lodestone, and showing that a magnetic needle located on its surface would point to the opposite poles, an experiment with which he confirmed his grandiose assumption that the earth is a big magnet. He showed that many other bodies, in addition to those already known, could be electrified by friction, and he introduced the adjective "electric" for those phenomena. He then distinguished between electric and magnetic phenomena by showing, for example, that a lodestone requires no stimulus to show magnetic properties, while glass and amber need to be rubbed. He also demonstrated that magnetic attraction is not impeded by a sheet of paper, but electric attraction is. He even had some understanding of the broken magnet experiment, according to which, by breaking a magnet into two parts one obtains two new complete magnets, each with two poles, and it is *not* possible to obtain two fragments, one with a north pole only and the other with a south pole only. Galileo knew some of Gilbert's books, and spoke with admiration of his ideas concerning the earth as a giant magnet. Gilbert was less fortunate when he tried to theorize on those subjects, introducing suitable effluvia or emanations coming from electrified bodies. For more than two hundred years, electricians devised fluids, emanations, effluvia, and so forth without greatly advancing the theoretical understanding of electricity. But on the whole, Gilbert's pioneer work on electricity and magnetism is of the first rank. He discovered new phenomena and put forward new ideas. More than fifty years after Gilbert's death, Newton was well aware that electric and magnetic phenomena were important but little understood, as he says at the end of the *Principia* and in the interesting Query 22 of his *Opticks*.

The eighteenth century is populated with students of electricity who made important single discoveries. Certainly, to find that there are conductors and insulators of electricity is a greater discovery than many for which our contemporaries became famous, but in two hundred years who will remember many of our present leading physicists? Time is a severe judge, and only very few survive its corrosive action. For almost two hundred years, electricity and magnetism were advanced by the collective work of many distinguished physicists who are now forgotten by all except the professional historians. No towering figure exists, but, on the whole, towering work was done. We have to arrive at the time of Alessandro Volta (1745–1827), two hundred years after the publication of *De magnete,* to find concentrated revolutionary progress. However, to understand it, we must pass through the fertile valley of eighteenth-century electricity.

The first advance from the rubbing of glass and amber was the making of machines that would perform the rubbing efficiently. Otto von Guericke (1602–1686), burgermeister of Magdeburg and scion of an illustrious family, spent most of his time in political affairs concerning his home town. The Thirty Years' War was raging in Germany, and Guericke, who was trained as an engineer, had ample opportunity to use his professional skills, as well as his ability as a diplomat in favor of his town. When peace came, he found leisure to study the nature of the vacuum. He built a pump—the first vacuum pump—with which he performed many spectacular experiments, starting from about 1654. His contemporaries were stunned by the inability of sixteen horses to separate two evacuated hemispheres. Guericke's main goal was to experiment on the behavior of phenomena in a vacuum, such as would exist in interplanetary space. He conceived of many "virtues" that

Otto von Guericke's electrostatic machine is shown at the far right of the picture. At the center a feather floats, repulsed by the electrified sulphur sphere.

would act at a distance, such as gravity, as well as an expulsive virtue that expels fire and keeps the moon at a distance, the impulsive virtue (inertia), and virtues that cause light, sound, and heat. In order to exhibit some of them, he built a sphere of sulphur mixed with various minerals that could be electrified by friction. It showed several virtues and was the first electrostatic machine, too.

The next major progress in the study of electricity, the discovery of conductors and insulators, is due in large part to the Englishman Stephen Gray (1666–1736). He was born in Canterbury, the son of a dyer. Although he followed his father's trade, he also obtained a fair education, possibly in London. He came in contact with Astronomer Royal John Flamsteed (1646–1719) and performed good astronomical observations, which procured him an invitation from Roger Cotes (1682–1716), who wanted to set up an observatory at Trinity College. Gray remained at Cambridge about one year, and then returned to his birthplace. In 1711 he petitioned for admission as a pensioner to Charterhouse in London, an institution devoted to the support of impoverished gentlemen.

It was there that Gray made his fundamental discovery. It came in a rather roundabout way. He started experimenting with a long glass tube, electrified at one end and closed at both ends with cork stoppers. Chance observations led him to modify the experiment by inserting in one of the corks a stick directed outside the tube, and he noted that the electrification imparted on the glass propagated to the cork and the stick. Later he extended the experiments to great distances. Suspending

One of Stephen Gray's many experiments investigating electric conduction. Here he demonstrates the conductivity of a body. (Johann Gabriel Doppelmayr, Neu-entdeckte Phaenomena von bewündernswurdigen Würckungen der Natur, *Nuremberg, 1744.)*

a string with silk threads, he found that it could carry electricity for over 300 feet. However, the silk threads broke under the weight, and when he replaced them with stronger metal wires, he found that he could no longer transmit his electrical effects. He finally interpreted this and other experiments by distinguishing insulators from conductors.

Such discoveries were far from easy, especially if one has at his disposal only a glass tube or a primitive electrostatic machine. The observations of conduction, induction, electrification, and the role of the earth as a body (as we would now say) at a constant potential are all intertwined. Thus, for example, a charge can be transmitted from an isolated conductor to another by conduction through a conducting wire, but it can also appear by induction, by simply bringing the conductor near to the charged body. The disentangling of all these effects is clearly difficult, even more so as strings and twines may be conductors or insulators, depending on their composition, the humidity of the air, and many other variables that are controllable only once they are known to be important. The investigations are full of pitfalls, and it takes uncommon ingenuity and critical sense to reach the correct answers. In James Clerk Maxwell's *Treatise on Electricity and Magnetism,* written in 1873, the experimental foundations of electrostatics are reduced to seven experiments: electrification by friction; electrification by induction; electrification by conduction; and four more that distinguish vitreous from resinous electricity (positive from negative charge) and give methods of measuring charges. J. J. Thomson (1856–1940), in editing the third edition of the *Treatise,* adds a footnote: "The difficulties which would have to be overcome to make several of the preceding experiments conclusive are so great as to be almost insurmountable." The eighteenth-century electricians collectively overcame them.

The next important electrician is Charles Dufay (1698–1739), a Frenchman. He came from a prominent family, and his father had high connections in the army and in the Church. He secured for his twenty-five-year-old son an appointment as adjunct chemist in the Académie des Sciences. The young Dufay had no scientific credentials, but the appointment was felicitous because he soon showed uncommon ability. In 1732 he was appointed superintendent of the Royal Gardens, and his methodical electric studies started in 1733. He found that there were two kinds of electricity, and only two. He called them vitreous and resinous because they were liberated by rubbing glass or rosin, respectively; they are the positive and negative charges of today. It is remarkable that it took more than a hundred years from the time of Gilbert to establish this fundamental fact: Dufay found that equal electricities repel each other and different ones attract each other. In 1733 Dufay took a collaborator for his electrical studies: Jean Antoine Nollet, who became the famous Abbé Nollet (1700–1770). Nollet was the son of a peasant; his illiterate father had agreed to have him educated by the Church, and Nollet attended school but did not become a priest. He had also developed uncommon manual ability and established relations with several French savants, among them René-Antoine Réaumur (1683–1757) and Dufay. In time he became prominent, was appointed tutor to the royal heirs, and ensured his fame by his six-volume treatise on physics, *Leçons de physique expérimentale,* which remained for many years a standard text. He developed a theory of electricity that prevailed for some years, but has no interest to physicists today.

A technical advance of the first magnitude was brought about by the invention of the electrical condensor. By chance observation, it was found that by charging a liquid in a bottle and holding the bottle in the hand, one could obtain powerful discharges. The fact was observed in 1745 by E. G. von Kleist (*ca.* 1700–1748) and publicized by Petrus van Musschenbroek (1692–1761) of Leyden. Musschenbroek was a known physicist, author of a treatise that became one of the standard texts of his time. The apparatus that accumulated or condensed electricity became known as the Leyden jar. It was soon found that the presence of the liquid was unnecessary and that it could be replaced by a conducting foil covering the inside of the bottle. Years later, in 1753, John Canton (1718–1772), under the inspiration of Benjamin Franklin's (1706–1790) theories, was led to another important observation: A conductor brought near a charged body, but not touching it, shows an electric charge of opposite sign to that of the charged body in the region near it and a charge of the same sign in the region far from it. This effect, called electrical induction, is radically different from other methods of electrification.

By the middle of the eighteenth century, electrical experiment had become very fashionable. It was common to electrify people by insulating them and connecting them to an electrostatic machine. These demonstrations were performed in social gatherings in elegant salons, as well as to a paying larger public; they even reached colonial America.

One of those demonstrations had the momentous consequence of arousing the interest of an American spectator, Benjamin Franklin. Franklin was born in Boston, halfway in time between the Pilgrims and the Boston Tea Party, the son of

An electrified lady sets spirits on fire by means of an electric spark. The phenomena of electricity were quite fashionable at the time. (From Treatise on Electricity, *an anonymous work, Venice, 1746.)*

Benjamin Franklin (1706– 1790), the American printer and statesman, who was also distinguished for his electrical studies that found practical application in the lightning rod. Franklin is possibly the most distinguished figure of the American Enlightenment in the colonial era.

an English soap and tallow maker. His delightful autobiography gives a vivid account of his life and times. He was to become in many ways a representative of the American ideal and the voice of America to Europe. His father had left England for religious reasons; in New England, he married and had a total of fourteen children from two wives. The children entered various trades, and Benjamin, who had shown marked learning ability, went to grammar school, but was apprenticed when he was twelve years old to his older brother, a printer. Benjamin very early wrote newspaper articles — amusing and original (*Dogood Papers,* 1722). However, he did not get along with his brother, and he moved from Boston to Philadelphia, where he established himself as a typographer and bookseller. By 1724 he had moved to London, where he continued working as a printer and became acquainted with, among others, Henry Pemberton (1694– 1771), the editor of the third edition of Newton's *Principia.* Franklin hoped to meet Newton through him, but it did not happen. Instead, he met a lesser light, Peter Collinson (*ca.* 1693– 1768), a fellow of the Royal Society who was to be his lifelong friend and important for his later electrical work.

Returning to Philadelphia in 1726, Franklin continued his work as a typographer, but soon branched out into journalism, publishing, first, the *Pennsylvania Gazette* and, in 1733, the famous *Poor Richard's Almanac.* His fame started to spread, he received official appointments, such as postmaster, and grew influential in Pennsylvania politics. In due course, he also became rich. He had a strong penchant for practical invention: Franklin stoves replaced old fireplaces, improving fuel economy, and are still in use. By the time he had reached the age of forty, he could afford to devote himself to what he really liked without financial concern. In 1743 he printed the "Proposal for Promoting Useful Knowledge Among the British Plantations in America," which in due course was implemented by the creation of the American Philosophical Society.

The following are Franklin's words on his acquaintance with electricity:

> In 1746, being at Boston, I met there with a Mr. Spence, who was lately arrived from Scotland, and show'd me some electric experiments. They were imperfectly performed, as he was not very expert; but being on a subject quite new to me, they equally surprised and pleased me. Soon after my return to Philadelphia, our Library Company received from Mr. P. Collinson, Fellow of the Royal Society of London, a present of a glass tube, with some account of the use of it in making such experiments. I eagerly seized the opportunity of repeating what I had seen in Boston; and, by much practice, acquired great readiness in performing those, also, which we had an account of from England, adding a number of new ones. I say much practice, for my house was continually full, for some time, with people who came to see these new wonders.

From that time on, Franklin experimented in electricity and developed his own ideas on the subject. He communicated his results in letters, mostly to Collinson, who printed them as a book, first published in 1751. The book was a great success and went through many editions and translations. Franklin's scientific contributions led to his election as a foreign member of the Royal Society of London in 1756. Political activity was, however, becoming more and more absorbing, and, in Franklin's mind, had to take precedence, as a matter of civic duty. He was in England from 1757 to 1762 as an agent of the Assembly of Pennsylvania, and after he had helped draft the Declaration of Independence in 1776, he returned to Paris for nine years on diplomatic assignments. He returned to Philadelphia in 1785 and later served as a member of the Constitutional Convention. He died in Philadelphia in 1790 at the age of eighty-four.

Franklin was a representative of the Enlightenment in an American context, a sort of homespun Voltaire, simpler and less sophisticated than the French philosopher. He also had appeal as being nearer to the state of Nature celebrated by Rousseau. He became immensely popular in Europe. He had left his narrow, sectarian Quaker religion early in his life and become a deist of liberal opinion. His morals were practical and utilitarian, and so was his science, often accompanied by technical invention.

His most important accomplishment from a theoretical point of view was the novel use of inferences from the principle of the conservation of charge. This principle had been perceived by several independent investigators, for instance, William Watson (1715–1787), but Franklin exploited it. In his view, a body contained equal amounts of positive and negative electricity, which, under normal conditions, exactly neutralize each other. Electrification is separation of the two electricities, which may be called positive and negative, with the implication that their sum remains constant and is zero. Franklin demonstrated those ideas by experiments in which two people, standing on insulated platforms, took the electricity from a glass tube rubbed with a cloth. One of the experimenters took it from the glass, the other from the cloth. When their fingers came near each other, a spark flew from one to the other, and both were neutralized. Similar experiments varied in many ways, but not in principle.

Important as that result was, Franklin's fame with the public depended much more on his experiments in atmospheric electricity, culminating with the invention of the lightning rod. At the time, the ideas concerning fire, combustion, lightning, sparks, and electrical discharges were far from clear. Franklin surmised that lightning was a gigantic electric spark. He had shown that a body with a sharp point will easily lose its electric charge, and by combining the two ideas, he thought of gradually discharging a building, and in that way protecting it from a sudden thunderbolt:

> I say, if these things are so, may not the knowledge of this power of points be of use to mankind in preserving houses, churches, ships, etc., from the stroke of lightning, by directing us to fix on the highest parts of these edifices, upright rods of iron made sharp as a needle, and gilt to prevent rusting, and from the foot of those rods a wire down the outside of the building into the ground, or down round one of the shrouds of a ship, and down her side till it reaches the water? Would not these pointed rods probably draw the electrical fire silently out of a cloud before it came nigh enough to strike, and thereby secure us from that most sudden and terrible mischief?

Experiments performed by Franklin and others, first in France and later in Philadelphia, demonstrated that one could indeed draw electricity from the clouds. Franklin's famous experiment with the kite passed without mishap, but the physicist Georg Wilhelm Richmann (1711–1753) was killed in a similar one in Russia. Until then the experimenters did not realize the great danger of their attempts.

Joseph Priestley (1733–1804), an Englishman who later emigrated to the United States, is best known as one of the great chemists of his age because of the discovery of oxygen. He was a friend of Franklin and an electrician. In 1767 he wrote a book entitled *History and Present State of Electricity* without suspecting that electricity had barely begun. In the book he reported an experiment that had already been performed by Franklin and which he confirmed. It showed that in a closed metallic box, there is no electric force inside and no charge on the inner surface. Priestley knew his Newton and inferred that this experiment could be explained by assuming that electric charges of equal sign repel each other with a force proportional to the inverse square of their distance. The deduction is correct, but it was not pursued further for a while, and it did not attract the attention it deserved.

By approximately 1770 the phenomenology of electrostatics was established. It was known that there were either two electricities, positive and negative, or maybe only one, but such that it could be added or drawn from a neutral body. Electricity was conserved, i.e., the sum of the positive and negative charges was constant. There were insulators in which electricity could not move, and conductors in which it moved freely. Equal charges repelled each other, while opposites attracted each other. With these fundamental facts understood, the time was ripe for establishing a quantitative law for attraction and repulsion. The Newtonian precedent of gravitation must have been present to many minds.

The tragic end of an electrical experiment performed by G. W. Richmann (1711–1753) in St. Petersburg. The investigator approached an isolated lightning rod during a storm and was killed; his assistant was knocked unconscious. Richmann was a distinguished scientist of German origin.

Direct experimental proof of the inverse-square law was first obtained by John Robison (1739–1805), an adventurous Scotsman who built an ingenious apparatus with which he measured the distance dependence of the electric force. He did not publish his results, however, for many years. In the meantime, Charles Augustin Coulomb (1736–1806) had nailed down the force law, which is now justly called Coulomb's law. Coulomb was a mathematically well-trained French engineer. His father was an army officer turned tax collector, who had hoped to make his son a doctor. The father lost his fortune, and the son, who was interested in science and engineering, managed to enroll in the military engineering school of Mézières. He spent nine years in tropical Martinique, rebuilding forts destroyed in the Seven Years' War, and returned to France when he was thirty-six years old and in poor health. He shared a prize offered by the Académie des Sciences for the study of magnetic variations. This brought him into the Académie in 1781 and, more important, to the invention and analysis of the torsion balance, an instrument of incomparable sensitivity. He used this tool extensively in electric and magnetic investigations, culminating in the demonstration of the inverse-square law which he plainly proclaimed in a memoir of 1788. Coulomb's result was accepted immediately in France, but much more slowly elsewhere, although in England it had been recognized, but not published, by that country's leading electrician, Henry Cavendish (1731–1810).

Cavendish was the scion of a great family and an extreme eccentric who had little contact with his fellowmen. Cavendish's father, Lord Charles Cavendish, had already done valuable work in electricity. The son, and heir to one of the great English fortunes, was born in Nice, where his mother was living for health reasons. She died when Henry was only two years old. The young Cavendish attended Cambridge University at Peterhouse, but, as was then common for aristocrats, he did not graduate. After the traditional tour on the continent, he settled in London in his father's home. Cavendish was an extraordinary experimenter and a major physicist, but his eccentricities were equally extraordinary. He dressed differently from everyone else, spoke confusedly and hardly intelligibly, and shunned society and especially women. His relations with other scientists were reduced to a minimum, even though he became a Fellow of the Royal Society in 1760 and a member of the French Institute in 1803. He lived extremely parsimoniously, although he financed all his experimental work. At his death he left over a million pounds, a fortune comparable to that of the great modern American foundations.

Cavendish's claim to fame is more as a chemist than as a physicist because his chemical discoveries were published whereas only a small fraction of his electrical work was revealed by him in a difficult paper published in 1771. Most of his results came to light through the efforts of Maxwell, who, as Cavendish professor, undertook the publication of Cavendish's electrical work. William Thomson (1824–1907), had come across some measurements by Cavendish of the capacity of condensors of different shapes. The results were surprisingly precise, and he suggested further examination of Cavendish's papers.

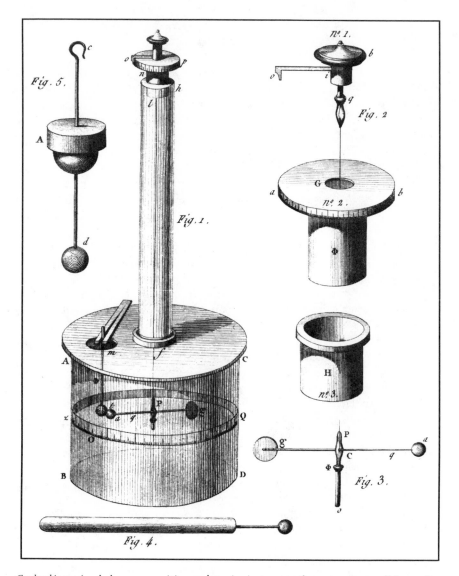

Coulomb's torsion balance, a sensitive and precise instrument for measuring small forces, from
Construction et usage d'une balance electrique *(1785). The torsion of the wire, which is*
measured by the knob on the tube, compensates the attraction or repulsion between charged bodies.
(Courtesy of the Lawrence Berkeley Technical Information Center, University of California,
Berkeley.)

  In the unpublished work, one found the proof of the inverse-square law by
the absence of the electric field inside of a charged conductor, using Newton's
mathematical arguments. Cavendish's definition of capacity of a conductor followed
from his principle that conductors charged ''in the same degree'' (we would now say

at the same potential) contain amounts of charge proportional to their capacities. The amount of charge may be measured directly by discharging the object with a test probe. He also found that two condensors having the same geometry have a capacity that depends on the dielectric. Finally, he measured the resistance of various bodies — for example, he said that water saturated with salt is 560,000 times more resistant to the passage of electricity than iron, having the same geometrical configuration. We can admire all these discoveries, but they were as inaccessible as the diamonds that undoubtedly exist in the mines.

In the published work of Cavendish, we find the determination of the attraction between two heavy spheres. He used the torsion balance for this purpose. The result gives the gravitational constant γ that completes Newton's inverse-square law: $f = \gamma mm'/r^2$. By calculating with this constant the acceleration of gravity g, we find ρ, the mean density of the earth, because $g = (4/3)\pi\gamma\rho R$. Here R is the earth's radius.

Help from the Frogs: Galvani and Volta and "The Most Marvelous Instrument Invented by Mankind"

With the phenomenology of electrostatics in hand, and with the knowledge of the inverse-square law, a complete mathematical description of the phenomena of electrostatics, inspired by the Newtonian idea of action at a distance became possible. The great mathematical physicists in the time of Napoleon or shortly thereafter — Pierre-Simon Laplace (1749–1827) and Siméon-Denis Poisson (1781–1840) in France, George Green (1793–1841) in England, Carl Friedrich Gauss (1777–1855) in Germany, and others — developed that theory in a form that has stood up to the present. Joseph Louis Lagrange (1736–1813) had introduced, for gravitation, the fundamental concept of "potential" (1772); Laplace had found the equation for the potential in a vacuum (1782); and Poisson had extended the Laplace equation to the case in which there are charges present (1813). The differential operator "Delta," or Laplacian, was to become one of the mainstays of mathematical physics. Green and Gauss discovered fundamental properties of the potential for the Newtonian case, contained in the famous formulas that carry their names.

Magnetism followed the fortunes of electricity, except that free magnetic charges were not found — only doublets of equal amounts of positive and negative magnetism. Otherwise, it was not difficult to extend the mathematics of electrostatics to magnetostatics.

Thus, a big part of electricity was reaching a stage of maturity and, one could even think, completeness, when major experimental discoveries opened up new horizons and revealed that what one knew was even less than the proverbial tip of the iceberg.

The new departures came from a completely unexpected quarter — the work of a professional anatomist and biologist. Luigi Galvani (1737–1798) of

Bologna. He was a well-known professor who had taught anatomy and obstetrics in his native town. Electricians had for many years found physiological effects of electric discharges and connections, real or imaginary, between electrical and biological phenomena. Much of this work was in error, some was fraudulent, and, on the whole, the subject did not elicit much respect. Galvani, on the other hand, was highly esteemed, and what he reported in a Latin publication of 1791 entitled (in translation) *Commentary on the Forces of Electricity in Muscular Motions,* attracted immediate attention and serious consideration from the profession.

Galvani had started his experiments a few years earlier, but here are his words in free translation from his publication of 1791: "I dissected and prepared a frog in the usual way, and with other things in mind, I placed it on a table not far from an electrical machine. When my assistant touched with a scalpel the crural nerves there was a strong contraction of all the muscles of the leg. Another person present remarked that he believed the thing happened when the electrical machine gave a spark. . . ." Puzzled by the observation, Galvani "was pushed with incredible alacrity to find out what was going on and to bring to light what secret was hiding behind the phenomena." He varied the experiments in many ways. He found that atmospheric electricity acted on the frogs, and that if one puts metallic foils on the muscles, so as to make a sort of Leyden jar, with the frog as the bottle, the contractions increase. A big effect is also obtained by touching the nerves and the legs with a metallic arc, and the effect is much bigger if the arc is composed of two different metals. By these experiments Galvani hoped to find out the nature of the animal spirit that had been sought for so long.

Luigi Galvani (1737–1798), professor of anatomy at the University of Bologna. His good work in medical disciplines has been overshadowed by his startling discoveries of electrical effects in dead frogs. Those discoveries opened a new era in the study of electricity.

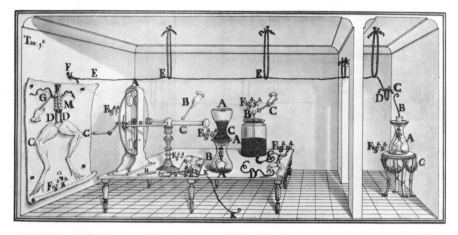

Galvani's laboratory, as portrayed in an original drawing that was used for etchings illustrating his work. Note the frogs prepared for experiments, the electrostatic machine, the Leyden flask, and the conductors. (Courtesy of the University of California, Berkeley.)

Galvani had caught a much bigger fish than he had thought; he had opened two great chapters of science — electrophysiology and the study of electric currents. That he could not disentangle the problem is natural. Nobody could for many years, and Galvani became involved in a scientific controversy, chiefly with Volta, who saw clearly one part of the problem, while the other is not completely clarified even now. Galvani died in 1798, still convinced that animal electricity was not the same as ordinary electricity.

We meet now for the first time, Alessandro Volta, after whom the "volt" is named. He was born in 1745 in Como, Italy, to a well-to-do Catholic family. His father had been a novice in the Jesuits for eleven years before marrying a noblewoman, who was also very religious. Volta's father had three brothers in holy orders, and of his nine children, five joined the Church. Alessandro always had great respect for his brother, the Archdeacon, and for his best friend, Canon Giulio Cesare Gattoni, but Alessandro, after having been educated by the Jesuits, preferred a secular life, although the clerical society surrounding him was, on the whole, rather merry, ready to enjoy life, and reasonably open-minded. Volta cohabited for many years with a singer, but married somebody else, only when he was about fifty. The wife has been described as a homely woman, very noble, rich and wise.

Volta's education had centered on Latin, languages, and literary subjects. He could write occasional sonnets in French and Italian and odes in Latin. His scientific inclinations seem to have grown spontaneously. By the age of nineteen, he wrote a small poem in Latin hexameters on chemical discoveries. The region around Como where he lived was prosperous, communications with Switzerland were easy and

Alessandro Volta (1745–1827), professor of physics at the University of Pavia and one of the greatest experimental electricians. He invented the voltaic pile (that is, the electric battery), which started modern electricity. (Courtesy of the Biblioteca Comunale di Como.)

immediate, the ruling Austrian government was definitely liberal for the time, and rich people of the region enjoyed an easygoing, refined life.

The young Volta started experimenting in electricity, read what books he could find, and became deeply interested in that work. His friend Gattoni supported him by giving him some apparatus and space in his house. By the age of sixteen, Volta had started to write letters to several known electricians, including the Abbé Nollet in Paris and Giambatista Beccaria (1716–1781) in Turin. Beccaria was a well-established, internationally known electrician, who admonished Volta to make fewer theories and base himself more on experiment. In fact, Volta's theoretical ideas in his youth were less important than his experiments. As the years passed, Volta came to understand static electricity at least as well as the best electricians of his time. Soon he began creating original instruments that applied his theories. The nub of it, in modern terms, was that he had gained a clear idea of quantity of electricity, potential or tension, as he called it, and electric capacity as well as of the relation $Q = CV$.

An excellent example of a Volta instrument is the electrophorus. A conducting plate is first laid on top of a charged "pie" of rosin that has been electrified by friction. The metal plate is then grounded by touching it and, afterward, lifted, using an insulating handle. The plate is thus charged at a high potential and can be used to charge a Leyden flask. The operation can be repeated indefinitely. The contrivance is most ingenious and was later developed into a whole family of electrostatic machines. Volta felt strongly that he had to measure quantitatively his electrical magnitudes, and he thus devised an electrometer, the forerunner of all absolute electrostatic electrometers, that could measure potential differences in a reproducible way. He established a scale for his instrument, and from his description, we can reconstruct that his unit was 13,350 of our volts. The invention of the electrophorus led to Volta's being given a job as professor of physics in the schools of Como (1775). His fame began to extend beyond Italy, and the Physics Society of Zurich elected him as a member.

Volta's interests were not confined to electricity. He discovered methane by observing it bubbling in marshes near the Lago Maggiore. Combining his chemical and electrical interests, he built an apparatus called an eudiometer, in which he could burn gases in a closed vessel, igniting them by electric sparks. At the age of thirty-two, he took a trip to Switzerland, during which he met Voltaire and several of the Swiss physicists. On his return, he was appointed professor of physics at the University of Pavia, the leading university of Lombardy. He kept that post until his retirement, and it was there that he made his most epochal findings.

Volta made another long trip abroad in 1792; this time he did not limit himself to nearby Switzerland, but went to Germany, Holland, France, and England, visiting, and occasionally experimenting with, his most important colleagues, such as Laplace and Antoine-Laurent Lavoisier (1743–1794). He then was elected a corresponding member of the French Academy and, not much later, a foreign member of London's Royal Society.

Shortly after his forty-fifth birthday, Volta read Galvani's papers of 1791, which were to set him up for his greatest inventions and discoveries. He was at first skeptical, but soon started experimenting on things that, in Volta's words, "were superior to anything that was known up to then about electricity, so much that they seemed prodigious." Initially, he agreed with Galvani's conception of the frog as a Leyden flask, but after a few months, he began to suspect that the frog is chiefly a detector and that the source of the electricity is outside the animal. He noted also that if two different metals in contact with each other are placed on the tongue, one feels a special sensation, sometimes acid, sometimes alkaline. He assumed, and could prove by electrostatic measurements that elicit our admiration, that two different metals, such as copper and zinc, acquire different potentials when in contact. He measured this potential difference, obtaining results not too different from what we now know is the contact potential difference between them. Galvani's experiments were thus explained, at least when the metallic arc connecting the muscles with the nerves is bimetallic, by assuming that the frog is simply an extremely sensitive electrometer. Of course, Galvani answered that he could observe the contractions even when the metallic arc was of one metal only. That was a serious objection, and Volta pointed to inhomogeneity in the metal and other causes for his defense.

A deeper study of the problem by Volta led to the invention of the pile, one of the marvels of the ages. Volta found that conductors of electricity could be divided in two classes. The first class contained metals that obtained different potentials when in contact, and the second class contained liquids (electrolytes, in modern language) that did not acquire a potential very different from a metal immersed in them. Furthermore, conductors of the second class, in contact, do not acquire appreciably different potentials. Conductors of the first class could be ordered in a scale so that each was positive with respect to the following — for example, zinc with respect to copper. In a chain of metals, the difference of potential between the first and the last is the same as if the intermediate contacts did not exist, and the first and last members of the chain were in direct contact.

Volta ultimately arrived at the idea of combining a number of conductors of the first and second kind in such a way that the potential differences generated in each contact would add. He called the instrument "pile" because it was composed of a pile of discs of zinc, copper, and cloth soaked with an acid, repeated many times. He wrote of his invention in a famous letter (in French) to Sir Joseph Banks (1743–1820), president of the Royal Society, under the title "On the electricity excited by the mere contact of conducting substances of different kinds."

The pile generated a continuous electric current of an intensity greater by order of magnitudes from that obtainable by electrostatic machines, and thus started a true revolution in science. Dominique François Arago (1786–1853), writing in 1831, echoes some of the admiration: ". . . This pile of couples of different metals separated by a little liquid, is, as far as the singularity of the effects it produces, the most marvelous instrument ever invented by mankind." He went on to describe what was known at the time, and we must remember that in 1831 there were not yet important practical applications of the electric current.

Volta's greatest achievement had come when he was relatively old, at age fifty-five. It was immediately acclaimed by all physicists. In 1801 he went to Paris and demonstrated his experiments to the Académie Française in the presence of Napoleon, who requested that Volta be given a special gold medal and a pension. Volta continued his life as a protégé of Napoleon, just as he had been a protégé of the Austrian Emperor Joseph II twenty years earlier. When he asked to retire from the Pavia chair in 1804, Napoleon refused his request, gave him more honors and money, and made him a count. After the downfall of Napoleon, Volta accommodated himself without much trouble to the returned Austrians. He thus passed through that agitated period of history unscathed, honored by whomever was in power, while he seemed profoundly indifferent to politics and concerned only with his studies.

After the pile work, Volta practically disappeared from the scene. The exploitation of his discovery fell to others. He may have been too old to compete with younger, fresher forces; he may even have been psychologically hampered by the magnitude of his previous achievements. He did not leave a school; his way of working was possibly too personal, and the absence of formal mathematics in his writings and teachings may have restricted his ability to communicate. Volta spent the last eight years of his life pretty much in seclusion between his villa at Camnago and nearby Como. He died on March 5, 1827 at the age of eighty-two.

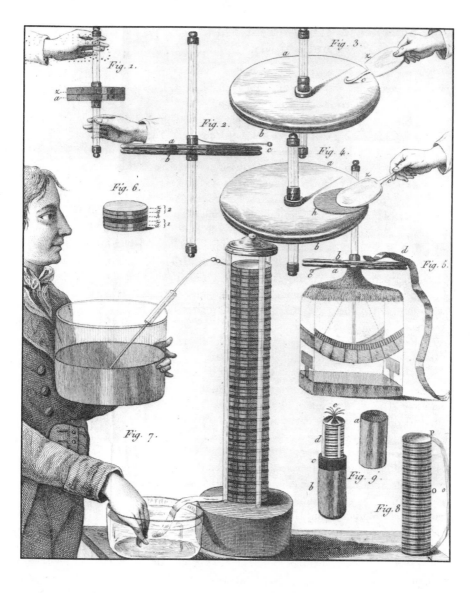

In three lectures to the Académie des Sciences in November 1801, Volta summarized his electrical work. This illustration is taken from the Bulletin of the Philomatic Society, *a private association of distinguished scientists that reported the lectures. The experiments show the potential difference that arises on contact between different metals. Fig. 5 depicts the condensing electroscope used for the measurements. Figs. 7, 8, and 9 show various forms of the pile and demonstrate how it is used to charge a condenser. (Courtesy of the University of California, Berkeley.)*

Electromagnetism: The Current and the Needle — Oersted and Ampère

With the invention of the pile and the availability of strong electric currents, the study of electricity acquired a new dimension.

Electrostatic machines could give high voltages, perhaps 30,000 volts, judging from the length of the sparks they delivered. The energy liberated in the discharge epended on the condensors complementing the machine — 1 joule per discharge is a reasonable estimate — but, of course, the discharges occurred in a certain time interval because the condensers had to be recharged. One watt is the order of magnitude of the average power delivered. With voltaic piles, the voltage depended only on the number of units in the pile, and a little over 1 volt per unit is the voltage obtained. Well-financed laboratories had batteries of hundreds of units. In Paris, Napoleon gave the École Polytechnique a battery of 600 units; in England, Sir Humphry Davy (1778 – 1829) had one of 3,000. The current that such batteries delivered depended only on the resistance of the circuit, including the battery itself, but currents of 10 amps were not exceptional. Thus, the power delivered could be of the order of 10 kilowatts, 10,000 times greater than that obtained from an electrostatic machine. The differences in the effects observed was such that for many years physicists doubted whether the electricity of the electrostatic machine and that of the voltaic battery (often called galvanic current) were of the same nature.

Very quickly, physicists and chemists explored the consequences of the passage of electric current in various substances, and a host of results enriched chemistry. The phenomena of electrolysis provided the possibility of decomposing salts and discovering new chemical elements. William Nicholson (1753 – 1815), having heard of Volta's invention before its publication, decomposed water by passing a current through it, and observed the evolution of oxygen and hydrogen. Davy, of which more will be said later, showed that soda and potash contained two new metals, sodium and potassium. The voltaic arc was also soon discovered and studied. Thus, the extraordinary effects of the electric current revealed new posibilities and kept physicists and chemists busy.

A new field of immense consequence opened up at the end of July 1820 with the sensational announcement that an electric current deflected a magnetic needle initially oriented parallel to it. The news came from a professor of physics at the University of Copenhagen, Hans Christian Oersted (1777 – 1851). The son of a pharmacist in a small Danish town, at age twelve Hans was sufficiently educated to help his father in the shop, a task which stimulated his scientific interest. He studied medicine, physics, and astronomy at the University of Copenhagen, and by the turn of the century, when all Europe was in ferment, he started his professional life as an apothecary in Copenhagen. As soon as he heard of Volta's discovery, he began experimenting with electric currents, and in 1801 he embarked on a traditional European tour that took him to France, Germany, and the Netherlands. In Germany he met the famous philosophers Frederick Wilhelm Schelling (1775 – 1854) and

Hans Christian Oersted (1777–1851), professor of physics at Copenhagen, discovered in 1820 the magnetic action of electric currents, opening up the study of electromagnetism. (Courtesy of the Technical University of Denmark.)

Johann Gottlieb Fichte (1762–1814), as well as the physicist Benjamin Thompson (Count Rumford, 1753–1814) and other scientists. By 1803 he had returned to Copenhagen and applied for a professorship of physics at the university. He obtained it three years later. That advancement put him in the higher social classes of his country, but he does not seem to have done anything of permanent value in physics for many years.

Oersted had philosophical ideas that vaguely advocated the unity of the forces of nature, and he was inclined to romantic conceptions on the completeness of nature. On the whole, I have the impression that these nebulous German ideas were of very modest value, and were possibly even an impediment to a modern understanding of science. However, when he observed the effect of the current on the magnetic needle, he immediately perceived that he had made a great discovery. He was then giving a course on electricity, and preparing for it. He says:

> These lectures and the preparatory reflections led me on to deeper investigations than those which are admissible in ordinary lectures. Thus my former conviction of the identity of electric and magnetic forces developed with new clarity, and I resolved to test my opinion by experiment. The preparations for this were made on a day in which I had to give a lecture in the evening. I there showed Canton's experiment on the influence of chemical effects on the magnetic state of iron. I called attention to the variations of the magnetic needle during a thunderstorm, and at the same time I set forth the conjecture that an electric discharge could act on a magnetic needle placed outside the galvanic circuit. I then resolved to make the experiment. Since I expected

the greatest effect from a discharge associated with incandescence, I inserted in the circuit a very fine platinum wire above the place where the needle was located. The effect was certainly unmistakable, but still it seemed to me so confused that I postponed further investigation to a time when I hoped to have more leisure. At the beginning of July (1820) these experiments were resumed and continued until I arrived at the results which have been published.

This account, written by Oersted early in 1821, has the ring of authenticity and sets forth his greatness and his limitations.

The scientific publication of his discovery took the form of a Latin paper, ''Experiments on the effect of the electric conflict [current] on the magnetic needle.'' In it he carefully describes his experiments and notes that whatever deviates the magnetic needle is not absorbed by conductors or insulators, and that the action of the current is observed only on magnetic substances. The last sentence of the paper reads, ''I shall merely add to the above that which I have demonstrated in a book published seven years ago that heat and light consists of the conflict of the electricities. From the observations now stated, we may conclude that a circular action likewise occurs in these effects. This I think will contribute very much to illustrate the phenomena called polarization of light.''

A lithograph of André Marie Ampère (1775–1836), the ''Newton of electricity,'' according to Maxwell. He laid the theoretical foundations of electromagnetism.

Oersted's paper was mainly qualitative, but it opened the gate to the study of electromagnetism. Its importance was immediately recognized, and the paper was translated into German, French, and English and published in standard scientific journals. It was communicated to the Académie des Sciences on September 11, 1820, by Arago. Among those present was André-Marie Ampère (1775 – 1836), an able mathematician equipped with all the tools of modern analysis, which he taught at the École Polytechnique. In one week Ampère succeeded in giving a complete quantitative theory of Oersted's observation and in laying the foundation of a mathematical theory of electromagnetism.

A vivid description of those days transpires in a letter dated September 25, 1820, addressed by Ampère to his son, then nineteen years old, who was traveling in Switzerland.

> Every moment has been taken by an event important in my life. From the very moment I have heard for the first time of the beautiful discovery of M. Oersted at Copenhagen concerning the action of a current on the magnetic needle, I have continuously thought about it. I have written a great theory of these phenomena and of all others known for magnets and tried the experiments indicated by this theory. All have succeeded and have revealed to me new facts. I read the beginning of a Memoir last Monday. In the following days I performed confirmatory experiments some time with Fresnel, some time with Despretz. I repeated them all on Friday at Poisson's house. . . . Everything succeeded marvellously, but the decisive experiment that I had thought as final proof demanded two galvanic piles. When I tried it at my house with Fresnel with piles that were too weak, it did not succeed. Yesterday I obtained permission from Dulong for Dumotier to sell me the great pile that was under construction for the Physics course at the Faculty and the experiment has been done at the house of Dumotier with full success and repeated today at 4 o'clock at the séance of the Institut. No objections were raised this time and here we have a new theory of the magnet that reduces all its manifestations to electric currents. This does not agree at all with previous opinions. . . .

Ampère perfected his theory in a period of a few months. He kept refining it for some years, and in 1827 his great work appeared, *Memoir on the Mathematical Theory of Electrodynamical Phenomena, Derived Solely from Experiment,* in which he summarized his experiments. He put at its foundation four experiments. He showed, first, that two equal currents in opposite directions, near to each other, have no action on a magnetic needle, and, second, that no action obtains if one of the two wires is bent and crooked with a number of small sinuosities. The third experiment shows that a conductor, capable of moving only in the direction of its length and carrying a current that enters and leaves at fixed points of space, is unaffected by any closed current placed in its neighborhood. The fourth experiment uses three circular currents of radii R', R'', R'''. The coplanar circles have their centers O', O'', O''', on a straight line and the radii of the circuits are such that $R'/R'' = R''/R'''$ and equal also to the ratio between the distances $O'O''$ and $O''O'''$. The same current goes clockwise through the three circles or clockwise in the external circles and counterclockwise in the central one. In those conditions, the central circle is in mechanical equilibrium when the external circles are fixed.

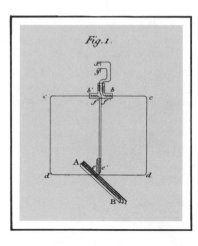

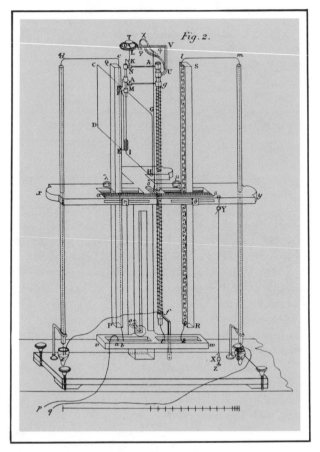

Ampère's four experiments by which he established the action between two elements of electric current. Fig. 1 shows the general form of the apparatus, insensitive to terrestrial magnetism. Fig. 2 portrays its practical realization and the equivalence of the effect of a straight and a sinusoidal current. Fig. 3 demonstrates the absence of forces parallel to the current. The apparatus in Fig. 4 shows that the force between two elements of current varies as the inverse square of their distance. (From André Marie Ampère, Mémoire sur la théorie mathématique des phénomènes électrodynamiques, uniquement déduite de l'expérience, Paris, 1827.)

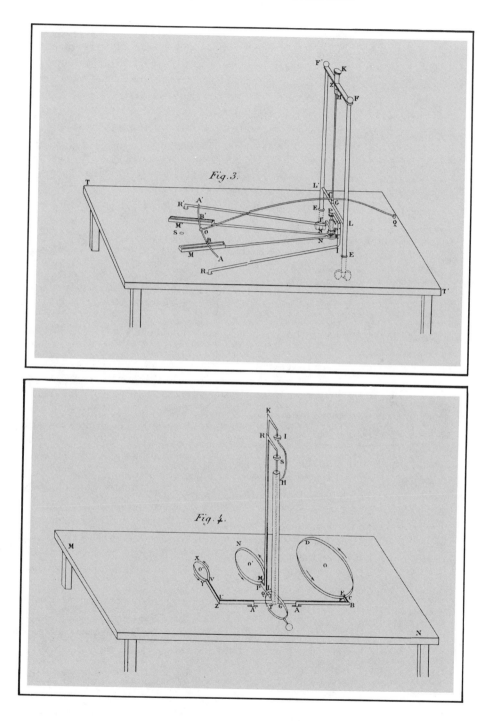

From these four experiments Ampère derived a law of force between two elements of wire through which currents flow. He gave a rather complicated expression for this force, but with its help, he could account for all observations. He showed that outside of its body, a permanent magnet was exactly equivalent in its magnetic action to a cylindrical coil of wire carrying a current — a solenoid, as he said. He explained the magnetism of a substance such as iron, assuming that each molecule contains a permanent closed current. Magnetization was the orientation of all the molecules under the influence of an external field. That view was truly prophetic; it contained many of the ideas by which we now explain magnetism. Augustin Jean Fresnel (1788 – 1827) also contributed some important points of view to the development of Ampère's ideas.

Maxwell's opinion on Ampère's work given in his *Treatise on Electricity* is:

> The experimental investigation by which Ampère established the laws of the mechanical action between electric currents is one of the most brilliant achievements in science. The whole theory and experiment seems as if it had leaped, full grown and full armed, from the brain of the "Newton of electricity." It is perfect in form and unassailable in accuracy, and it is summed up in a formula from which all the phenomena may be deduced, and which must always remain the cardinal formula of electrodynamics.

After that praise, however, Maxwell made some remarks, to which we shall return later.

Ampère the "Newton of electricity," was similar to Newton in that he, too, was neurotic. His father was a prosperous merchant in Poléymieux, a small village near Lyons in France. Ampère remained deeply attached to the places of his infancy. Many times in correspondence with his closest friends, he intimated that if he could go back to Lyons, all would be well with him. Ampère never attended a regular public school, but he acquired an immense store of knowledge from his readings. As a child he showed uncommon intelligence, exceptional arithmetic ability, and a prodigious memory. He avidly read what he found at home and as an adult could, without effort, repeat long articles on such odd subjects as heraldry that he had read in the French Encyclopedia. As a very young boy, he asked the librarian of the public library of Lyons for the works of Euler and the Bernoullis. The librarian was somewhat startled, and said that in addition to their difficult content, they were also written in Latin. The lad was taken by surprise by the language, but in a few weeks, he had learned Latin and read the mathematics, too. Very early he also devised a universal language, with a strong logical base.

The French Revolution almost destroyed Ampère by executing his innocent father. For more than a year, Ampère was devastated. Slowly his interest in nature and his deep religious faith helped him recover his balance. Nevertheless, all his life he remained a character, the quintessential absentminded professor and, at a deeper level, the victim of severe depressions.

The young Ampère made firm, lifelong friendships with several devoutly Catholic young men of Lyons. The friends had common religious and cultural interests, and, when separated, they corresponded for many years. Many of the letters are preserved, and they afford insight into the intimate life of Ampère, his unhappiness, his doubts, his religion, and occasionally his enthusiasm.

On an August evening in 1796, Ampère was collecting plants for his botanical studies when he suddenly met a group of girls. He fell in love with one of them at first sight. A diary, interspersed with poetry, that he kept at the time informs us of the events. The impecunious pair was soon engaged, but could afford to marry only three years later. They were extremely happy; soon they had a child, Jean-Jacques, who later became a well-known literary figure in France, as well as a steady correspondent with his father. Ampère earned a living by teaching in provincial schools. When he could return to Lyons he felt he had reached all he wanted, but the years of happiness were cut short by his wife's death in 1803.

Ampère's fame rose rapidly, and he obtained a succession of jobs. In 1808 he was appointed inspector of the French University, an office he retained for life. By 1814 he was a member of the Académie and a professor. As an inspector, his official duties included inspections of a great number of schools, examinations, and attending administrative functions for which he was unsuited. When he returned to Paris, and in his successive chairs of mathematics, physics, and philosophy, he would mention his theorem of Avignon, theory of Marseilles, and so on, from the places where he had invented them.

In later years Ampère became quite lazy in reading; he who as a child had devoured the French Encyclopedia, now shrank before the task of reading memoirs on physics that had been submitted to him. He would comment, "You do not know how lazy I am."

Ampère's scientific work can be divided into three periods. The first was devoted to pure mathematics. The second, from 1808 to 1815, was devoted to chemistry, sometimes in collaboration with Joseph Louis Gay-Lussac (1778–1850). He worked on the halogens and on chemical theory. He was a convinced atomist and arrived at Amedeo Avogadro's (1776–1856) rule later than Avogadro but independently. The third period, from 1820 to 1827, was devoted to studies on electromagnetism, where he achieved his most important discoveries. Later, he turned to philosophy and to the classification of sciences (coining the word *cybernetics*).

Ampère's personal life after his first wife's death was stormy. In 1806 he remarried, but it was a disastrous decision—the marriage broke apart with great bitterness on both sides, leaving a daughter who later married an undesirable husband. Ampère's second mother-in-law helped to exacerbate the situation, too. Ampère's mother and an aunt came to help him in raising his family, but he had to struggle continuously with financial difficulties, in part caused by his innocence, his credulity, and his family problems. For all his genius, he would believe impostors who demonstrated to him experiments on "animal magnetism." The combination of his famous absentmindedness, shortsightedness, and physical clumsiness gave rise to many stories, such as the one that he erased a blackboard with his handkerchief and immediately thereafter used it to wipe his face, or of his not recognizing his colleague Napoleon Bonaparte at the Institut de France.

The range of his interests was formidable. He could debate with Georges Cuvier (1769–1832) on the philogeny of animals, showing outstanding expertise as a naturalist, and the chairs he held in succession impressively confirm his universality. He was a professor of mathematics at the École Polytechnique from 1809 to 1819, professor of philosophy at the Faculty of Letters in the Paris University from 1819, professor of astronomy in 1820, and professor of physics at the Collège de France from 1824 until his death. He also composed poetry and was very sensitive to simple music but elaborate music seemed to disturb him. Arago, who knew and loved him well, had to say, "When one speaks of contemporary scientists whose immense gifts have been misused, the name of Ampère is the first that occurs."

A Fresh Look and a Powerful Imagination: Faraday — From Bookbinder to Prince of Science and Experimenter Supreme

Ampère's discoveries, as well as those of other electricians of his time, such as Poisson, were couched in a perfect mathematical language, representing the ultimate development of Newtonian physics. The basis was always to find the law of force between two elements of current or two point charges in motion. Many physicists labored on this problem, including Wilhelm Weber (1804–1891), Franz Neumann (1798–1895), Carl Neumann (1832–1925), Ludwig Lorenz (1829–1891), and others. They found formulas describing all the facts of electrostatics and of electrodynamics that were supported by known experiments. These formulas contained a universal constant that turned out to be equal to the velocity of light. Ultimately, however, that line of endeavor was superseded by Maxwell's theory.

Although Newton had expressed doubt about the idea of action at a distance, most subsequent physics, as well as Newton's, was based on that idea. I personally was never disturbed by the idea of action at a distance; however, this idea is objectionable to many. With what we know now about atoms, action at contact in a gross mechanical way is a complex manifestation of Coulomb's force, a typical action at a distance. On the other hand, those who wanted to introduce a medium came to grief with the ether and its mechanical properties, which are certainly objectionable.

Be all this as it may, there is no doubt that ascribing a fundamental role to the medium through which electrical forces propagate has revolutionized electrical science and has advanced it tremendously. That progress, to a large extent, is due to Michael Faraday (1791–1867) and Maxwell, who brought electricity to its classical culmination. Practical applications are not the object of this book, but we cannot be blind to the fact that electricity has changed our mode of life and has generated a vast field of engineering applications. For many of them — electric motors, generators, and transformers — the electricity of actions at a distance is sufficient. However, electric waves and radio communications require for their explanation an understanding of Maxwell's theory. The ether, or its substitutes, is essential for them.

The final outcome of the study of electricity along the new lines, stressing the part of the medium, is embodied in the famous *Treatise on Electricity and Magnetism* by Maxwell, which appeared first in 1873. A quote from the preface to this book well represents the views of the mature Maxwell:

The general complexion of the treatise differs considerably from that of several excellent electrical works, published, most of them, in Germany, and it may appear that scant justice is done to the speculations of several eminent electricians and mathematicians. One reason of this is that before I began the study of electricity I resolved to read no mathematics on the subject till I had first read through Faraday's *Experimental Researches on Electricity*. I was aware that there was supposed to be a difference between Faraday's way of conceiving phenomena and that of the mathematicians, so that neither he nor they were satisfied with each other's language. I had also the conviction that this discrepancy did not arise from either party being wrong. I was first convinced of this by Sir William Thomson, to whose advice and assistance, as well as to his published papers, I owe most of what I have learned on the subject.

As I proceeded with the study of Faraday, I perceived that his method of conceiving the phenomena was also a mathematical one, though not exhibited in the conventional form of mathematical symbols. I also found that these methods were capable of being expressed in the ordinary mathematical forms, and thus compared with those of the professed mathematicians.

For instance, Faraday, in his mind's eye, saw lines of force traversing all space where the mathematicians saw centres of force attracting at a distance: Faraday saw a medium where they saw nothing but distance: Faraday sought the seat of the phenomena in real actions going on in the medium, they were satisfied that they had found it in a power of action at a distance impressed on the electric fluids.

When I had translated what I considered to be Faraday's ideas into a mathematical form, I found that in general the results of the two methods coincided, so that the same phenomena were accounted for, and the same laws of action deduced by both methods, but that Faraday's methods resembled those in which we begin with the whole and arrive at the parts by analysis, while the ordinary mathematical methods were founded on the principle of beginning with the parts and building up the whole by synthesis.

I also found that several of the most fertile methods of research discovered by the mathematicians could be expressed much better in terms of ideas derived from Faraday than in their original form.

The introduction of electric and magnetic fields in many cases produces a great simplification in computation. For instance, we can look at the problem of the force between two circuits by calculating the field produced by one of them, and then by calculating the force due to the field acting on each element of the second circuit. The result is, of course, the same as obtained by Ampère's formula, but the calculation is simpler and clearer (see Appendix 4). Hence, we must start with a presentation of Faraday.

One day in the winter of 1812, when Napoleon was losing his army on the Russian plains, a young man of twenty-one presented himself at the Royal Institution of London and requested an interview with its famous director, Humphry Davy. As a recommendation, he brought a book containing notes he had taken at Davy's lectures. The book was neat and well bound, and the boy made a good impression on Sir Humphry, who happened to have a vacancy for an assistant. After some time, he

Sir Humphry Davy (1778–1829) was a brilliant chemist, among the first to apply electrolysis to the discovery of elements (Na, K, Ba, Sr, Ca, Mg). For a period he was the director of the Royal Institution, where he hired Faraday. He became famous for the invention of the safety lamp for miners and was elected president of the Royal Society in 1820.

hired the applicant, who turned out to be Michael Faraday, one of the greatest physicists of all time. As we shall see, it did not take very long for Davy to become aware of that fact.

Faraday was born on September 22, 1791, in a village that now is part of London. He came from a family of manual workers, with no special culture and rather poor. Michael's father was a blacksmith. The child received an indifferent schooling, and when he was thirteen years old, he was apprenticed to a bookseller, stationer, and book binder. Besides binding the books, the apprentice read them. His master encouraged him, and one of the customers gave him tickets to Davy's lectures. Faraday recognized his vocation, which must have been powerful, and soon he was part of the personnel of the Royal Institution. In 1813 Sir Humphry and Lady Davy decided to go on a tour of the continent, and they took Faraday with them as a secretary, but also for more menial tasks. The trip lasted eighteen months and was fundamental to Faraday's education. For one thing, he met with many famous scientists such as Ampère, Volta, Arago, and Gay-Lussac, and some of them recognized the value of the modest youth accompanying Davy. He became a lifetime friend of Gustave De La Rive and his son Arthur-Auguste (1801–1873), distinguished physicists of a Geneva family that was intellectually and politically prominent in the life of the Canton, and well beyond it. I mention La Rive here because, to an extent, he acted as a clearinghouse for ideas and remained a constant correspondent with Faraday.

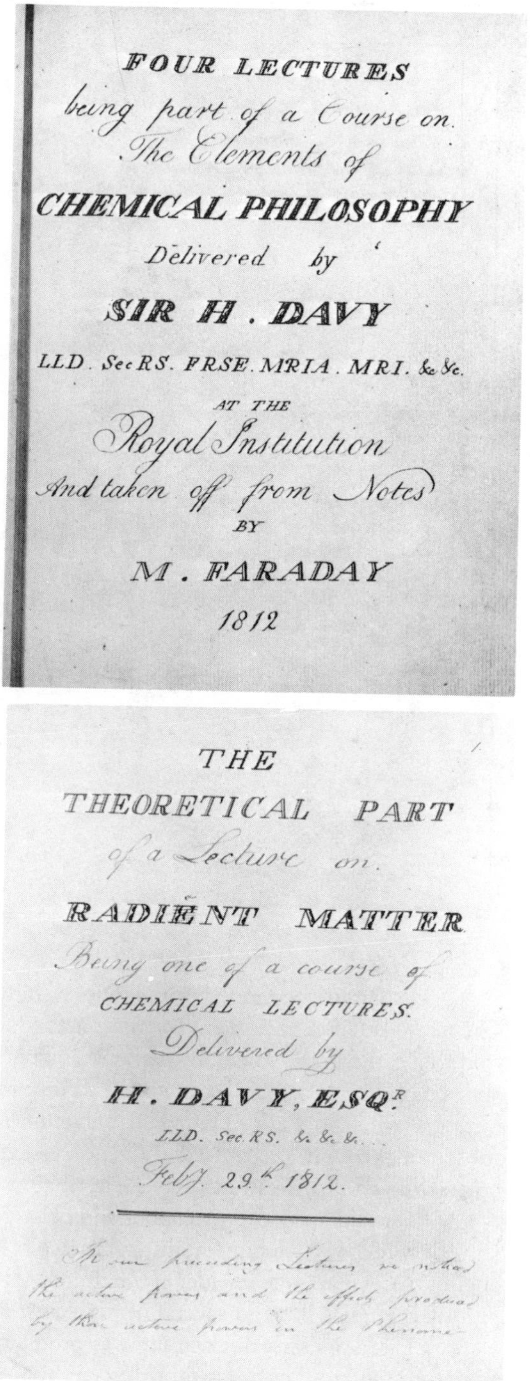

FOUR LECTURES
being part of a Course on
The Elements of
CHEMICAL PHILOSOPHY
Delivered by
SIR H. DAVY
LLD. Sec RS. FRSE. MRIA. MRI. &c &c.
AT THE
Royal Institution
And taken off from Notes
BY
M. FARADAY
1812

THE
THEORETICAL PART
of a Lecture on
RADIENT MATTER
Being one of a course of
CHEMICAL LECTURES.
Delivered by
H. DAVY, ESQ.R
LLD. Sec RS. &c &c &c...
Feby. 29th. 1812.

In our preceding Lectures we noticed
the active powers and the effects produced
by those active powers in the Phenome-

Title pages of Faraday's notes of Sir Humphry Davy's lectures in 1812. These notes were Faraday's own recommendation to gain employment at the Royal Institution of London. (By courtesy of the Royal Institution, London.)

A drawing of young Faraday. (By courtesy of the Royal Institution, London.)

Davy's trip extended from France to Italy (down to Naples), to Switzerland, Germany, and Belgium. Faraday was an assiduous correspondent. He had friends from his early youth to whom he kept writing, in what seems to me a somewhat verbose style, moralizing occasionally, but more often lively and effective. For instance, when in Paris, Davy collaborated with Gay-Lussac in studying a new substance, which he recognized as the new element iodine. That dramatic event and others are vividly told in Faraday's letters.

Faraday was deeply religious and belonged to a small sect, the Sandemanian Church, which was fundamentalist in character. He remained faithful to it all his life, and for many years was an elder of the church. Apparently, religion was an important part of Faraday's life and several of his closest friends came from the same sect.

Faraday's scientific activity was prodigious. For a few years after his return from the grand tour of Europe, he devoted his time to practical chemical analysis and to his duties as a subordinate at the Royal Institution, including important help to Davy. His first published paper, dated 1816, is on the nature of caustic lime in Tuscany. Many years later Faraday reprinted it in his works and noted, "Sir Humphry Davy gave me the analysis to make as a first attempt in chemistry at a time when my fear was greater than my confidence, and both far greater than my knowledge; at a time when I had no thought of ever writing an original paper on science."

By the end of his career, around 1860, Faraday's laboratory notes contained more than sixteen thousand entries carefully numbered in sequence and bound in

A daguerrotype of Faraday with J. F. Daniell, the inventor of a much-used type of unpolarizable battery. (By courtesy of the Royal Institution, London.)

volumes where the author delighted in showing his old proficiency as a bookbinder. Those notes, and several hundred more previous or posterior to the bound books, are edited and published in printed volumes, the most famous being his *Experimental Researches on Electricity*. The variety of subjects studied by the great "natural philosopher," as he called himself, is as follows, in chronological order: Research on alloys of steel (1818–1824); compounds of chlorine and carbon (1820); electromagnetic rotations (1821); liquefaction of gases (1823, 1845); optical glass (1825–1831); discovery of benzene (1825); electromagnetic induction (1831); the identity of electricity from various sources (1832); electrochemical decomposition (1832 on); electrostatics, dielectrics (1835 on); discharges in gases (1835 on); light electricity and magnetism (1845 on); diamagnetism (1845 on); "thoughts on ray vibrations" (1846 on); gravity and electricity (1849 on); time and magnetism (1857 on). A complete analysis of his colossal output is beyond the scope of this book and would require an inordinate amount of space.

The list of subjects shows that up to about 1830 Faraday was primarily a chemist, except for a short but important period following Oersted's discovery. In 1821 he first tried his hand at electricity and magnetism and possibly sowed the seeds that later matured in his great discoveries of the following decade. His first period of activity terminates with the year 1830. By then he had established himself as a very successful professional analyst and practical consultant, and, more important, he had acquired an international reputation based on his solid scientific achievements. Those included the preparation of several new carbon compounds, such as "carbon perchloride" in his nomenclature, or hexachloroethane, $CCl_3 \cdot CCl_3$, in modern parlance, and tetrachloroetylene, $CCl_2 : CCl_2$; as well as the study of the gas used for the lighting of London. (Faraday's brother worked in that industry.) The gas was produced by heating animal oil and stored in cylinders where it left a liquid. Faraday analyzed that residue with great care and ingenuity. He found in it a component having a fixed boiling point of $80\,^\circ C$ and the gross composition CH. It was benzene, one of the mainstays of organic chemistry. However, at the time of its discovery, Faraday did not realize its future importance, nor, of course, its peculiar molecular architecture. This list of chemical discoveries is not complete, but it shows that if Faraday had not done anything else, he would be considered a distinguished chemist.

In fact, in the twenties he also succeeded in liquefying several gases: the original apparatus he used was extremely simple — a sturdy glass tube bent as an inverted V. At one end he put the substance generating the gas; the other end he dipped in a refrigerating mixture. The evolving gas increased the pressure inside the tube. By such a simple artifice, he could liquefy chlorine, sulphur dioxide, hydrogen sulfide, carbon dioxide, nitrous oxide, ammonia, hydrogen chloride, and other substances. Oxygen, hydrogen, and nitrogen did not yield, and some of them acquired the name "permanent" gases. Only when the concept of critical temperature evolved and adequate technical means developed was it possible to liquefy all gases. The last, helium, yielded only to the efforts of Heike Kammerlingh-Onnes (1853–1926) in 1908.

Faraday at work in his laboratory. (By courtesy of the Royal Institution, London.)

Beginning in 1818, Faraday collaborated for several years with a surgeon, James Stodart, a fellow of the Royal Society, in trying to create improved steel — more rust resistant and taking a sharper edge than the available English products. Metallurgy at the time was still very much an empirical art: "wootz," produced in India, was the best blade steel known. Faraday and Stodart alloyed iron with several others metals — platinum, silver, palladium, chromium, and others — but Stodart died in 1823, and Faraday moved to other work. Possibly they could have found some capital results of modern metallurgy. Samples of their blades still exist, and some are of superior quality.

All those enterprises show the exceptional chemical and technological prowess of Faraday. He collected his experiences in a large book, more than six hundred pages, titled *Chemical Manipulation,* which he published in 1827. That was the only book, apart from his collected electrical researches and other research papers, that Faraday wrote. It had great success and remained a standard text for decades. Perusing it today, it gives an extraordinary impression of immediacy and freshness. The techniques and the instructions are, of course, those of Faraday's time, when there was no electricity and primitive gas distribution in the laboratory, and very few pure chemical products. On the other hand, one feels the true interest of the author for the pupil. He is verbose, but he points out the pitfalls, and, short of seeing him operate in the laboratory, one has a clear idea of his technical skills, his carefulness, and his love for experimental work. The power of his imagination and his genius cannot be exhibited, but many traits of his scientific personality are discernible.

A letter Hermann von Helmholtz (1821–1894) sent his wife in 1853 gives us further insight into Faraday's character:

> I succeeded in meeting the first physicist of England and Europe: Faraday. . . . This was for me a great and plesant moment. He is simple, gentle, and modest as a child. I have not yet met a man so likeable. Furthermore he was most obliging, in showing me personally all there was to be seen. This however was little, because some pieces of wood, some wire and some pieces of iron seemed to suffice him for making the greatest discoveries.

Faraday's base of operation was always the Royal Institution, with which he became identified. He lived there with his wife, Sarah (they had a happy marriage, but no children), until 1858, when Queen Victoria gave them the use of a royal

The front of the Royal Institution on Albermarle Street in London. (From a watercolor by T. Hosmer-Shepherd, ca. 1838, by courtesy of the Royal Institution, London.)

house. However, even then he kept his rooms at the Royal Institution, where he had his laboratories, perfectly suited to his needs. Even today, a visit to the Royal Institution makes you feel his spiritual presence, and one would not be surprised to find him at work in some corner of the building.

A look at the drawings showing him at work conveys a vivid impression of what must have been the extraordinary ability and eagerness of the artisan Faraday. I surmise that he must have physically enjoyed the use of his hands. He writes many times that he must experience new phenomena by repeating the experiments, and that reading is totally insufficient for him.

Davy wanted to show his appreciation of Faraday but had financial difficulties at the instiution. He proposed Faraday's appointment as director in 1825 as a sign of esteem. Shortly thereafter, Faraday established a regular series of "Friday Evening Discourses" that still persists. Faraday devoted much effort to perfecting his lecturing art and became famous for it. He formulated suggestions and guidelines complete to minute details, and these are still conveyed to present-day lecturers at the institution. As a practical result of his lecturing skill, although the tickets for the lectures at the Royal Institution were rather expensive, whenever Faraday lectured, the auditorium was full. Other lecturers on average filled two-thirds of the seats. In addition to the Friday lectures, Faraday instituted special popular lectures for children, held during the Christmas period. One of his Christmas lecture series, under the name "Chemical History of a Candle," has delighted and inspired countless young people (including this writer) for more than a century. It has been translated into many languages and is still in print.

In 1824 Faraday was elected a fellow of the Royal Society; he was thirty-three years old. Davy was president of the society and, touched by jealousy for his subordinate, whom he had so generously helped and protected, opposed the election. It was an unfortunate episode of human weakness. It should be noted that when Faraday applied for his first job at the Royal Institution, he said to Davy that he "wanted to escape from trade which I thought vicious and selfish and enter the service of science, which I imagined made its pursuers amiable and liberal. . . . [Davy] smiled at my notion of the superior moral feelings of philosophic men, and said he would leave to me the experience of a few years to set me right on that matter." (From a Faraday letter, 1829.)

Faraday's salary at the Royal Institution was only £100 a year, but in 1833 he was appointed Fuller Professor of Chemistry at the same institution, which doubled his salary. His professional income, however, was substantial, and he could have become a rich man, if this had mattered to him. In 1835 Lord Melbourne, then Prime Minister, offered him a pension of £300 from the Civil List. In so doing, he used some unfortunate words, and Faraday's pride was offended. He refused the pension in a harsh letter, and the Prime Minister had to apologize in order to induce Faraday to accept the grant. The story appeared in the newspapers and aroused considerable public interest.

As soon as it became possible for him, Faraday relinquished most of his professional work and severely curtailed his social life, devoting all his forces to experimental research. One has the impression that only experimental research really

Faraday delivering a Christmas lecture in 1856. In the audience were the Prince Consort Albert and the Prince of Wales (the future Edward VII), who later wrote Faraday a nice letter, thanking him for his teaching. (By courtesy of the Royal Institution, London.)

interested him. He did not participate in public activities and declined most honors offered him, including the presidency of the Royal Society, proffered to him in 1857. He had had disagreements with the powers of the society because he wanted to concentrate the elections only on scientific personalities, omitting members of the aristocracy or other important persons who were merely amateurs. In 1835 he stopped going to the meetings, although he communicated his scientific papers to the society.

Early in life Faraday started to complain about headaches, dizziness, and, above all, loss of memory. Those symptoms were attributed to overwork, and periods of rest produced temporary recovery. The most serious break-down happened in the years between 1839 and 1844, when he was forced to take a long vacation. In his diary of the period, when he was recuperating in Switzerland, his love for nature, plants, and animals is manifest, and his preserved herbary is perfectly mounted and ordered. He remarks that a forty-five-mile hike, nothing unusual for him, shows that his body is in not-too-bad shape; he would gladly trade his physical health for an improvement of his memory. His symptoms may have been due to mercury poisoning. Other chemists and physicists of the time suffered from similar troubles. Although he was incapacitated only at the beginning of the spell, his activity diminished for almost five years, only to resume in full in 1845.

The greatest period of Faraday's achievements lasted from 1830 to 1839, when he was the foremost contributor to the discovery of modern electricity. In 1821 he had investigated the action found by Oersted, and he made a great point that the magnetic action was at right angles to the direction of the current producing it. Also, Faraday succeeded in building a sort of electric motor, which shows the rotation of a wire in a constant magnetic field. He demonstrated the rotation even in the earth's magnetic field. The experiment made a deep impression on him and on his contemporaries.

We find in Faraday's writings a number of words that seem vague or imprecise to a modern; *action* and *power* are two examples. I believe that Faraday had something clear in his highly imaginative mind and that his powers of expression were not always adequate to explain what he meant. In fact, he had notions that closely parallel the modern idea of field. In some cases, his description becomes quite precise, and his lines of force visualized by the disposition of iron filings in the case of magnetic fields are perfect. But this is not always so, and if Maxwell had said what I reported on page 133, I believe he was overgenerous, or at least only a Maxwell could disentangle from Faraday's writings the most useful theoretical notions.

Faraday became convinced that the relation of electricity to magnetism had to be extended, and that if a current could produce a magnetic field, a magnetic field also had to be able to produce a current. This notion was by no means limited to Faraday. Certainly Ampère and Arago, among others, had similar ideas. Ampère made experiments that missed the discovery of electromagnetic induction by a hair, and Arago showed that a rapidly rotating copper disc under a magnetic needle sets it in rotation. The notion of a connection between electric current, motion, and magnetic field was dimly felt by other investigators. Faraday brooded over it for about ten years and made numeous experiments, all negative. By 1831 he was trying experiments to see whether a magnetic field threading a conducting loop could produce a current. The results were negative in the static case; it was the fifth time that he experimented on the problem. In the summer of that year, he built an iron ring on which he wrapped two coils of copper wire. He then noted that if he sent a current in one and connected the other to a galvanometer, the instrument would signal a current *not in the stationary state,* but only at the establishment or interruption of a current in the other coil. That was *the* clue he needed. By the end of September, he had developed a clear understanding and experimental demonstration of electromagnetic induction. He had grasped the vital point that to generate a current, a conductor had to cut the lines of magnetic force, one of his beloved conceptions. Once the nature of electromagnetic induction was understood, Faraday was able to explain Arago's observations and to invent an electromagnetic generator of currents—a primitive dynamo. In a few months of work, toward the end of 1831, he made tremendous progress, and in addition to epoch-making discoveries, he laid the foundations of the future electric industry. The story is told that when a politician asked him what good were his discoveries, he answered "At present I do not know, but one day you will be able to tax them." One can still see the primitive motors, generators, and transformers built by Faraday at the Royal Institution. They are the three fundamental pieces of the modern electric power industry.

Electricity has different manifestations and a very complex phenomenology; an electric current can be generated by the traditional electrostatic machines, by the voltaic cell, and by electromagnetic induction. What are the relations between electric currents generated in any of these ways? Are they all the same? The question was a serious one, and it had occurred to many before Faraday. The weight of evidence was in favor of a unique type, but there were no compelling studies and systematic investigations. Those were Faraday's next endeavor. He devised methods to measure what we now call quantity of electricity, including a ballistic galvanometer and a voltameter. The conclusion of the work was: "The chemical power, like the magnetic force, is in direct proportion to the absolute quantity of electricity which passes." Further studies on electrolysis led to the discovery of what are now called Faraday's laws, the most important being that "electrochemical equivalents coincide and are the same with ordinary chemical equivalents." That fact practically forced the conclusion that the carriers of electricity all had the same charge and masses proportional to the atomic weight, or, more precisely, to the atomic weight divided by the valence. The step from the atomic nature of matter to the atomic nature of electricity and to the electron was short, but Faraday did not take it. Faraday's own words are as follows. (The numbers at the beginning of these and other paragraphs written by Faraday reflect the numbering system he used in his diaries and laboratory journals.)

868. What, then, follows as a necessary consequence of the whole experiment? Why, this: that the chemical action upon 32.31 parts, or one equivalent of zinc, in this simple voltaic circle, was able to evolve such quantity of electricity in the form of a current, as, passing through water, should decompose 9 parts, or one equivalent of

The entry Faraday made in his laboratory journal on August 29, 1831, recording the discovery of electromagnetic induction. The vertical pencil trace probably means the material was used for publication in the Experimental Researches. *(By courtesy of the Royal Institution, London.)*

that substance: and considering the definite relations of electricity as developed in the preceding parts of the present paper, the results prove that the quantity of electricity which, being naturally associated with the particles of matter, gives them their combining power, is able, when thrown into a current, to separate those particles from their state of combination; or, in other words, that *the electricity which decomposes, and that which is evolved by the decomposition of, a certain quantity of matter, are alike.*

869. The harmony which this theory of the definite evolution and the equivalent definite action of electricity introduces into the associated theories of definite proportions and electrochemical affinity, is very great. According to it, the equivalent weights of bodies are simply those quantities of them which contain equal quantities of electricity, or have naturally equal electric powers; it being the ELECTRICITY which *determines* the equivalent number, *because* it determines the combining force. Or, if we adopt the atomic theory or phraseology, then the atoms of bodies which are equivalents to each other in their ordinary chemical action, have equal quantities of electricity naturally associated with them. But I must confess I am jealous of the term *atom;* for though it is very easy to talk of atoms, it is very difficult to form a clear idea of their nature, especially when compound bodies are under consideration.

The studies on electrolysis called for the creation of a number of words for concisely expressing concepts that would have required a longish description. Faraday considered himself ignorant in philology, not having had a classical education. He asked for help from a famous Cambridge don, the Reverend William Whewell (1794–1866), Master of Trinity College, philosopher and mathemati-

Some of the apparatus Faraday used in investigating electrolysis and gas discharges. (By courtesy of the Royal Institution, London.)

cian, who suggested many of our commonly used words on the subject: electrode, anode, cathode, ion, electrolysis, and so on.

Faraday had deep theoretical concerns about the explanation of his discoveries. He did not like actions at a distance. Newton, too, had had misgivings about them. However, in its subsequent evolution, Newtonian mathematical physics forgot those problems, and the inverse-square law became the dominant concept, obliterating the function of the medium. Faraday was no expert mathematician, and thus was possibly less impressed than others by the achievements of formal Newtonian theory. On the other hand, he had seen the lines of force as realized by iron filings near a magnet. He had also seen that they may be curved, and that for him was proof that the action would not always propagate in a straight line. Moreover, in his great discovery of electromagnetic induction, he had recognized that the central point was the cutting of the lines of force by the conductor. No wonder that these lines of force became the main guide to his thoughts. He also developed another concept — that of electrotonic state. It is not easy to translate the electrotonic state into valid modern terms, and, thus, I will not go further on it. Faraday tried repeatedly to explain his ideas, and I am not sure that they did not change substantially in time. The following is a description of the electrotonic state, written in 1838:

> 1729. It appears to me possible, therefore, and even probable, that magnetic action may be communicated to a distance by the action of the intervening particles, in a manner having a relation to the way in which the inductive forces of static electricity are transferred to a distance (1677.); the intervening particles assuming for the time more or less of a peculiar condition, which (though with a very imperfect idea) I have several times expressed by the term *electro-tonic state* (60. 242. 1114. 1661.). I hope it will not be understood that I hold the settled opinion that such is the case. I would rather in fact have proved the contrary, namely, that magnetic forces are quite independent of the matter intervening between the inductric and the inducteous bodies; but I cannot get over the difficulty presented by such substances as copper, silver, lead, gold, carbon, and even aqueous solutions (201. 213.), which though they are known to assume a peculiar state whilst intervening between the bodies acting and acted upon (1727.), no more interfere with the final result than those which have as yet had no peculiarity of condition discovered in them.

Maxwell identified the electrotonic state with the vector potential.

Ideas involving lines of force and fields, as we now say, extended naturally to electricity. Electric charges must modify the space between them, and Faraday reconsidered electrostatics on a basis different from the usual one. He confirmed in a famous demonstration of 1835, using a huge conducting cube, that the electricity was all on the surface of the conductor. He entered the cube and demonstrated that a charge conferred on the cube had no influence inside the cube. That is an experiment that can be used to prove the inverse-square law. Faraday was led by his notion of lines of force to meditate on the state of the insulators between charged bodies. He concluded that they must be in a sort of state of strain. To experiment on the subject, he took concentric conducting spheres and filled the space between them with different insulators. He discovered that the condensers thus formed had different capacities, although they were geometrically identical. He thus ascribed to the insulator a specific inductive power. He had been anticipated in this discovery by Cavendish, but no one knew of Cavendish's work because his manuscripts at the time were still unpublished.

Faraday explained the phenomenon qualitatively by introducing the concept of polarization of a dielectric. I prefer to report his own words because they give the flavor of his power and of his limitations:

> 1679. The particles of an insulating dielectric whilst under induction may be compared to a series of small magnetic needles, or more correctly still to a series of small insulated conductors. If the space round a charged globe were filled with a mixture of an insulating dielectric, as oil of turpentine or air, and small globular conductors, as shot, the latter being at a little distance from each other so as to be insulated, then these would in their condition and action exactly resemble what I consider to be the condition and action of the particles of the insulating dielectric itself (1337.). If the globe were charged, these little conductors would all be polar; if the globe were discharged, they would all return to their normal state, to be polarized again upon the recharging of the globe. The state developed by induction through such particles on a mass of conducting matter at a distance would be of the contrary kind, and exactly equal in amount to the force in the inductric globe. There would be a lateral diffusion of force (1224. 1297.), because each polarized sphere would be in an active or tense relation to all those contiguous to it, just as one magnet can affect two or more magnetic needles near it, and these again a still greater number beyond them. Hence would result the production of curved lines of inductive force if the inducteous body in such a mixed dielectric were an uninsulated metallic ball (1219. &c.) or other properly shaped mass. Such curved lines are the consequences of the two electric forces arranged as I have assumed them to be: and, that the inductive force can be directed in such curved lines is the strongest proof of the presence of the two powers and the polar condition of the dielectric particles.
>
> 1680. I think it is evident, that in the case stated, action at a distance can only result through an action of the contiguous conducting particles. There is no reason why the inductive body should polarize or affect *distant* conductors and leave those *near* it, namely the particles of the dielectric, unaffected: and everything in the form of fact and experiment with conducting masses or particles of a sensible size contradicts such a supposition.

While creating far-reaching new science, Faraday also lectured regularly and very successfully. In addition, in 1836 he became a member of the Senate of the University of London and scientific adviser to Trinity House. That last post required considerable experimental work in the laboratory and in the field in connection with lighthouses and other aids to navigation. He enjoyed that work and pursued it for thirty years.

His breakdown of 1840, as we have seen, extended until 1845. The early phase had been the worst. After a year or so, even if he did not resume active laboratory work, he was much improved, and he certainly resumed his meditations on electricity. One of Faraday's favorite ideas was that the various "Forces of Nature," such as electricity, magnetism, light, gravity, and possibly others influenced each other. He also spoke of the "Unity of Forces." Some of his speculations may appear completely metaphysical, except for the all-important fact that they led him to great discoveries.

Condensers with different insulators used by Faraday in the study of the dielectric constant. (By courtesy of the Royal Institution, London.)

In 1845 he decided to try the influence of various agents on light. The following are the introductory paragraphs of the report of that investigation, which communicate some of his guiding philosophy:

2146. I have long held an opinion, almost amounting to conviction, in common I believe with many other lovers of natural knowledge, that the various forms under which the forces of matter are made manifest have one common origin; or, in other words, are so directly related and mutually dependent, that they are convertible, as it were, one into another, and possess equivalents of power in their action. In modern times the proofs of their convertibility have been accumulated to a very considerable extent, and a commencement made of the determination of their equivalent forces.

2147. This strong persuasion extended to the powers of light, and led, on a former occasion, to many exertions, having for their object the discovery of the direct relation of light and electricity, and their mutual action in bodies subject jointly to their power; but the results were negative and were afterwards confirmed, in that respect, by Wartmann.

2148. These ineffectual exertions, and many others which were never published, could not remove my strong persuasion derived from philosophical considerations; and, therefore, I recently resumed the inquiry by experiment in a most strict and searching manner, and have at last succeeded in *magnetizing and electrifying a ray of light, and in illuminating a magnetic line of force.* These results, without entering into the detail of many unproductive experiments, I will describe as briefly and clearly as I can.

Polarized light, by then well known and essentially understood since the time of Fresnel, became one of Faraday's favorite subjects. He tried to see whether there could be a change of polarization by passing light through a piece of glass or crystal subject to an electric field. The results were negative, although John Kerr (1824–1907), with more refined experimental means, found in 1875 the effect Faraday had sought in vain. In September 1845 Faraday finally had a first great success. He passed linearly polarized light through a piece of a special kind of heavy glass prepared by him years earlier. A magnetic field was then excited with its lines of

Faraday's laboratory journal at entry 7504, September 13, 1845, recording the discovery of birefringence induced in glass by a magnetic field. At the end of the day, at entry 7536, is the remark, "Have got enough for to day." (By courtesy of the Royal Institution, London.)

force parallel to the direction of propagation. The plane of polarization rotated as long as the magnetic field was established. In his laboratory notebook he wrote: "There was an effect produced on the polarized light; and thus magnetic force and light were proved to have relations to each other. This fact will most likely prove exceedingly fertile, and of great value in the investigation of conditions of natural force. . . ."

Immediately Faraday made a number of experiments to check that the effect was genuine and disentangled correctly its main features. He then confided, "Have got enough for to day."

Diamagnetism was the next discovery. Most materials — the list of what Faraday tried includes a strange mixture of fifty substances, from metals, glass, and blood to water and wax — when formed as a needle orient themselves perpendicularly to the lines of force of a suitably inhomogeneous magnetic field. Furthermore, they are repelled from either pole of a magnet. This behavior is produced by very weak forces, very much smaller than those acting on iron in a magnetic field. The phenomena deserve close study, and Faraday devoted several months to the subject.

In 1846 Faraday was substituting for Charles Wheatstone (1802–1875) at the Royal Institution. Wheatstone was expected to report on some work of his, but had panicked at the last moment and fled the place. Faraday improvised a lecture, but it turned out that it fell short of the allotted time and the lecturer had to add some material on the spot. He turned to some speculations he had been mulling over in his head and presented them very guardedly. However, the word spread, and he ended up writing a short paper with the title "Thoughts on Ray Vibrations." The paper is not very clear, but it contains some startling and fundamental departures. At best, it could be called a clear premonition of the electromagnetic theory of light. This may be an exaggeration, but we must defer to Maxwell, who, eighteen years later, formulated the electromagnetic theory of light, and says:

> The conception of the propagation of transverse magnetic disturbances to the exclusion of normal ones is distinctly set forth by Professor Faraday in his "Thoughts on Ray Vibrations." *The electromagnetic theory of light, as proposed by him, is the same in substance as that which I have begun to develop in this paper,* except that in 1846 there were no data to calculate the velocity of propagation.

Moreover, notes by Faraday, only recently published, show that he indeed had a deep insight in the electromagnetic nature of light. Maxwell, in his generosity, was also just.

In 1852 Faraday published the twenty-ninth series of his *Experimental Researches in Electricity;* it was the last. It fittingly concludes:

> 3234. It would be a voluntary and unnecessary abandonment of most valuable aid, if an experimentalist, who chooses to consider magnetic power as represented by lines of magnetic force, were to deny himself the use of iron filings. By their employment he may make many conditions of the power, even in complicated cases, visible to the eye at once; may trace the varying direction of the lines of force and determine the relative

polarity; may observe in which direction the power is increasing or diminishing; and in complex systems may determine the neutral points or places where there is neither polarity nor power, even when they occur in the midst of powerful magnets. By their use probable results may be seen at once, and many a valuable suggestion gained for future leading experiments.

Lines of magnetic force indicated by iron filings. The figure comes from Faraday's laboratory journal, where he fixed the filings. Such images helped Faraday demonstrate and develop the field concept. (By courtesy of the Royal Institution, London.)

The elderly Michael Faraday with his wife, Sarah. (By courtesy of the Royal Institution, London.)

A typical Faraday instruction on how to obtain an image of a magnetic field, using iron filings, is shown in the illustration on page 152. All manipulative and technical details are minutely given.

In the 1850s Faraday's activity slackened. He suffered again from a lapse of memory that became progressively worse. He could still perform experiments, but not at his previous pace. He tried to find an interaction between gravity and electricity. The results were negative, but the search has gone from Faraday to Einstein to the present; it still goes on. In fallow discovery periods, Faraday tried to help the public, addressing himself to problems that were already plaguing London, such as the preservation of paintings from the damages of the polluted atmosphere.

In 1862 he performed his last experiment—the attempt to find the influence of a magnetic field on the light emitted by a source immersed in it. The result was negative because Faraday's instruments were inadequate to detect the minute effect. Thirty-four years later, Pieter Zeeman (1865–1943), then a young man directly inspired from reading of Faraday's attempt, repeated the experiment with superior instruments and discovered the Zeeman effect, one of the harbingers of the new atomic physics.

During his last years, Faraday was incapacitated by the progressive failure of his memory. In 1860 he gave his last Christmas lectures, and in 1865 he resigned the professorship at the Royal Institution. It was manifest that his health was faltering, and he thus resigned all his other positions, including, in 1864, the eldership of the Sandemanian Church. He died in 1867 at the age of seventy-six.

In commemorating him, Lord Kelvin who had known him well, said: "He had an indescribable quality of quickness and life. Something of the light of his genius irradiated his presence with a certain bright intelligence, and gave a singular charm to his manner which was felt by every one, surely, from the deepest philosopher to the simplest child who ever had the privilege of seeing him in his home — the Royal Institution."

Faraday was recognized both during his lifetime and by posterity as one of the greatest "natural philosophers" (as Faraday would have called his profession). Each epoch needs certain special qualities, and it is possible that Faraday's great success may be due in part to the time in which he was operating. What were his extraordinary qualities? A powerful imagination accompanied by great resourcefulness in experiment, a passion for work supported by corresponding stamina, a critical spirit that allowed him promptly to distinguish a spurious effect from a true discovery, and such open eyes that nothing escaped him. He also had some sound general ideas and made up for his mathematical ignorance by a deep geometrical and spacial insight, as well as by the capacity for sustained thought. Among notes left by him, there is the following:

> It puzzles me greatly to know what makes the successful philosopher. Is it industry and perseverance with a moderate proportion of good sense and intelligence? Is not a modest assurance or earnestness a requisite? Do not many fail because they look rather to the renown to be acquired than to the pure acquisition of knowledge, and the delight which the contented mind has in acquiring it for its own sake? I am sure I have seen many who would have been good and successful pursuers of science, and have gained themselves a high name, but that it was the name and the reward they were always looking forward to — the reward of the world's praise. In such there is always a shade of envy or regret over their minds, and I cannot imagine a man making discoveries in science under these feelings. As to Genius and its power, there may be cases; I suppose there are. I have looked long and often for a genius for our Laboratory, but have never found one. But I have seen many who would, I think, if they had submitted themselves to a sound self-applied discipline of mind, have become successful experimental Philosophers.

From those lines one can infer that there were no mirrors at the Royal Institution.

Faraday, although an excellent lecturer, had no direct scientific pupils or collaborators. His method of working and, even more, his way of thinking on physical subjects prevented him from founding a school. The lack of formal mathematical knowledge, coupled with his powerful and very fast imagination, made communication difficult. Only a Maxwell could really plumb his thoughts. Faraday himself says:

> I have never had any student or pupil under me to aid me with assistance; but always prepared and made my experiments with my own hands, working and thinking at the same time. I do not think I could work in company, or think aloud, or explain my

thoughts at the time. Sometimes I and my assistant have been in the Laboratory for hours and days together, he preparing some lecture apparatus or cleaning up, and scarcely a word has passed between us.

The assistant he mentions was a retired noncommissioned officer, Sergeant Anderson. He was the ideal helper for Faraday, and Faraday praises him highly in his papers. Anderson's unquestioning obedience is reflected in the following story, the authenticity of which I cannot guarantee. Early one morning, Faraday went to the laboratory and found his assistant stirring a melted mixture of some substance. Faraday was a little surprised and asked him what he was doing so early in the morning. The answer was, "You asked me to stir this mixture yesterday, and never told me to stop."

Faraday's way of doing physics was suited for a time in which there was an entirely new phenomenology to discover and in which the theoretical underpinnings were incomplete in the extreme. Unfortunately later, mediocre physicists went hunting for new phenomena, imitating, in their minds, Faraday. In my generation I have still seen them, trying to excuse their laziness in learning what was known, and wasting time and money on silly experiments; on the other hand, if Faraday had been alive, maybe the nonconservation of parity might have been discovered twenty years earlier.

Infallible in Physics: Maxwell

If Faraday was the greatest experimental physicist of the nineteenth century, James Clerk Maxwell was the greatest theoretician. He is the founder of modern electrical theory and one of the founders of thermodynamics and statistical mechanics. I will speak of him now, but we must remember that he is prominent also in the subject of

Glenlair, the family home of James Clerk Maxwell in southwest Scotland, as it appeared in 1880 with Maxwell's improvements of 1867. (Courtesy of C. W. F. Everitt.)

heat. William Thomson, later Lord Kelvin, one of the best physicists of the nineteenth century, also has such a dual role. Time, however, has enhanced Maxwell and perhaps slightly tarnished Kelvin. I will speak of him later in connection with what was possibly his most original work on heat.

James Clerk Maxwell was born in Edinburgh in 1831, the year of the discovery of electromagnetic induction, but he was raised at Glenlair, about sixty miles south of Glasgow, in a country house built by his father. The Clerks were a prominent Scottish family that had had several intellectuals in their lineage. They owned land that made them financially comfortable. Maxwell's father took the name "Maxwell" when he inherited a small estate in Dumfriesshire. At the time of Maxwell's youth, Glenlair was very isolated, and a pioneering atmosphere prevailed, stimulating Maxwell's father to technical, often odd, endeavors. Maxwell's mother, Frances Cay of Edinburgh, died of gastric cancer when James was only eight years old.

James was remembered by his cousins and friends as a very inquisitive boy who constantly asked, "What's the go of that?" and was not easily satisfied with the answer. Because he grew up in the country with peasant children as companions, he acquired a strong Scottish accent that he preserved all his life. After the mother's death, the relations between father and son became even tighter, and the boy accompanied his father constantly on the estate. Later letters show the great confidence and love that prevailed between the two.

When Maxwell was ten years old, he was sent to Edinburgh Academy. In the beginning he had a difficult time at school because of his accent and because he wore strange clothes designed by his father. He was at first rather solitary and acquired the nickname of "Dafty." He was absorbed in his own thoughts and not very sociable, although he had a keen sense of humor. In the middle of his school years, he became brilliant in all subjects, including English verse, which he composed throughout his lifetime.

At the age of fifteen, he submitted a paper on a geometrical way of tracing ovals to the Royal Society of Edinburgh. It was published. It is not an important paper, but it is remarkable for such a young author.

In 1847 Maxwell entered Edinburgh University. (It must be remembered that Scottish universities at that time usually were entered at a young age; they more nearly resembled secondary school than true universities.) Maxwell studied mathematics, philosophy, and physics. The physics professor gave him the run of the laboratory where Maxwell, who always loved apparatus and had considerable manual dexterity, tried his first experiments. In 1850 he moved to Peterhouse at Cambridge, a college favored by Scottish students. William Thomson was already a fellow of the College, and Peter G. Tait (1831–1901), the other member of the notable Scottish Trinity, was a student—senior to Maxwell.

By the time he entered Cambridge University, Maxwell was conversant with English literature and was himself an able poet. He had a vast knowledge of mathematics and physics, but he lacked order in his studies.

In his first years, his Cambridge career was not marked by any notable feature. He had to undertake systematic studies, although he had already written a

Maxwell's first cousin, Jemima Wedderburn, left a series of delightful drawings illustrating life at Glenlair. Top left, Maxwell as a small child; top right, Maxwell with his dog, Toby; middle right, Maxwell with his father and Toby; bottom, a hunting scene with family members. (Courtesy of C. W. F. Everitt.)

valuable paper on elasticity and a few others of lesser importance. His tutor, William Hopkins, noted his strong geometrical bent and his ability to find solutions to mathematical problems by synthetic methods. "It seemed impossible for him to think wrongly on any physical subject, but in analysis he seemed far more deficient," was Hopkins' judgment. Several of his contemporaries reported on his rapid and occasionally confusing conversation. It seems that his mind worked faster than that of any of the listeners, who however were pleased by his discourse on any subject. He was also chosen to a select club of twelve students, the "Apostles," who, in their own opinion, were the best men of the university. He attended the lectures of G. G. Stokes (1819–1903) the important mathematical physicist, and he again met Thomson. He had known him from Scotland, at least since 1850, when Maxwell attended the British Association Meeting in Edinburgh. Professor James Thomson, the mathematician father of William, was also present. Furthermore, Maxwell's father had consulted the elder Thomson on the choice of a college at Cambridge, and the two families knew each other well.

Maxwell prepared to compete for the tripos of 1854. The tripos are a famous written competitive examination in mathematics, very important for the future career of any Cambridge man. The first "wrangler" (the winner of the competition) was Edwin John Routh (1831–1907), a future famous coach at Cambridge and an original mathematician of some standing; Maxwell was second wrangler, as William Thomson had been years earlier. In the following competition for the Smith prize, Maxwell was first, bracketed with Routh. One of the questions on the examination was to prove the celebrated Stokes theorem, which is so important in Maxwell's electromagnetic theory. The Rev. Whewell, Master of Trinity — the same man who had been consulted by Faraday in connection with the electrolysis nomenclature — was one of the examiners. Those men had the feeling that Maxwell was the better mathematician, but his answers were less well organized.

Once freed from the grind required by the tripos, Maxwell started serious investigations on his own on two subjects — color theory and electricity. At the same time, he tried to obtain a fellowship at Trinity College that would solve his practical problems.

In discussing Maxwell's work, we must take into account his method of attacking problems. He meditated on a subject and wrote on it, then passed to another, maybe for a long period, returning to the first subject years later with new vistas and deeper insights. It is thus more appropriate to abandon a chronological order and to proceed by subjects. Between his most important works, he interspersed here and there investigations seemingly disconnected from his main themes, in which, however, he seldom fails to reach some noteworthy new result.

Maxwell started studies on color vision while still at Edinburgh, under the guidance of Professor James David Forbes (1809–1868), and continued them intermittently. Fundamental progress on color theory — a chapter of physiological optics — had been made by Thomas Young (1773–1829). The analysis of the sensation of color is quite complicated; it had started with Newton, who combined spectral colors, obtaining white and other colors; Young had arrived at the conception of three fundamental colors. Maxwell continued the investigation by studying

Maxwell at Cambridge in 1855, holding the color top of his first vision experiments.

with the help of a spinning top that allowed quantitatively measuring of the colors being mixed and made great strides in unraveling the mysteries of color vision. He returned to this subject repeatedly. He experimented on different individuals and studied the sensitivity of their retinas, detecting the influence of a yellow pigment that gives the sensation of the "Maxwell spot."

A singular product of these investigations is the first color picture, obtained by using photographic techniques in 1861. He projected it at the Royal Institution to an audience that included Faraday. The picture shows a Scottish tartan photographed through red, green, and blue filters and then projected through the same filters. Collodion plates were used, and in modern times it was at first difficult to understand how the experiment could have been successful, because such plates are completely insensitive to red. A hundred years later at the Kodak Research Laboratories the mystery was investigated by R. M. Evans, who found that the dyes used by Maxwell also reflected ultraviolet light that passed through his red filter. The red image had thus been obtained using ultraviolet light! A modern repetition of the experiment was completely successful.

Another important but isolated subject taken up by Maxwell is a study of Saturn's rings. It had been proposed as a theme for an Adams prize in 1855, and Maxwell, using all the resources of a refined analysis showed that to ensure stability, the rings had to be formed by loose materials. Solid bodies or fluid bodies would be unstable. The satellite Pioneer II confirmed the correctness of Maxwell's analysis by crossing the ring unharmed. Maxwell's analysis represents a real tour de force. It elicited general admiration and established Maxwell as one of the foremost mathematical physicists. Its greatest importance possibly was the direction it gave to Maxwell's thoughts toward problems that were to lead him to statistical mechanics.

We come now to Maxwell's two major achievements: his studies on electricity and his studies on the kinetic theory of gases. In both fields he made supreme contributions of lasting character.

Maxwell started studying electricity while in Cambridge in 1855. He had two spiritual mentors — Faraday and Thomson. We have seen the influence of the first, but William Thomson's was different. He was senior to Maxwell by a few years; he was at Cambridge when Maxwell entered and was already well-known. Thomson had great mathematical skill and a broad culture. He was, in particular, deeply steeped in Fourier's work, which he called to Maxwell's attention; his fertile brain produced one idea after the other, and he published many papers on a variety of subjects. He certainly had genius, but especially in his electrical work, he considered innumerable separate problems, giving clever and original solutions without arriving at the deep Maxwellian synthesis. Thomson's work was a mathematical inspiration for Maxwell and suggested many analogies to physical situations remote from the electrical field. Possibly the influence was stronger at the beginning of Maxwell's own work than later.

Maxwell's first large paper on electricity appeared in 1856. Its title was "On Faraday's Lines of Force." He gives a theory of the electric and magnetic field based on analogy. It is appropriate to quote here what he meant by analogy:

> In order to obtain physical ideas without adopting a physical theory we must make ourselves familiar with the existence of physical analogies. By a physical analogy I mean that partial similarity between the laws of one science and those of another which makes each of them illustrate the other. Thus all the mathematical sciences are founded on relations between physical laws and laws of numbers, so that the aim of exact sciences is to reduce the problems of nature to the determinations of quantities by operations with numbers. Passing from the most universal of all the analogies to a very partial one, we find the same resemblance in mathematical form between two different phenomena giving rise to a physical theory of light.

In the specific case of electricity, he notes the mathematical similarity between the potential equation and the equation of heat propagation in the stationary case. He then proceeds to analogies with hydrodynamics, and ultimately introduces the four vectors — E and H, which are forces, and B and I (current density), which are fluxes produced by the forces — and then establishes their relations and their equations. Later, to explain electromagnetism, he discusses the "electrotonic state" and connects it to a new vector A with the properties of the vector potential: $\mathbf{B} = \text{curl } \mathbf{A}$ and $\mathbf{E} = -\partial\mathbf{A}/\partial t$. He also makes many specific applications of this theory.

Five years later, in 1861, Maxwell returned to the subject and devised a medium that would behave elastically in such a way as to reproduce the electromagnetic forces. He did not mean to imply a literal interpretation, but simply a help to the imagination. The illustration below is taken from Maxwell's paper and shows the constitution of his model of space. From the properties of the model, Maxwell infers two momentous consequences: the usual conduction current has to be augmented by an amount proportional to $d\mathbf{E}/dt$, which he interprets as "general displacement of the electricity" and that the medium sustains transversal vibrations (not longitudinal), propagating with a velocity c that can be calculated from the laws of electricity. When he compared the numbers, "The velocity of transverse undulations in our hypothetical medium, calculated from the experiments of MM. Kohlrausch and Weber, agrees so exactly with the velocity of light calculated from optical experiments of M. Fizeau that *we can scarcely avoid the inference that light consists in the transverse undulations of the same medium which is the cause of electric and*

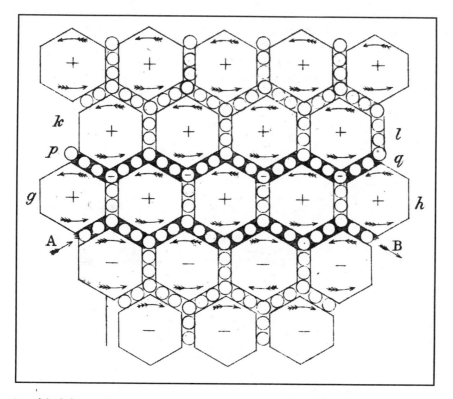

A model of electrical particles and vortexes in the ether used by Maxwell for explaining the properties of the electromagnetic field. There are some obvious errors in this figure, which the reader may easily discover. (From Maxwell's On Physical Lines of Force, *1861.)*

magnetic phenomena." The premonitions of Faraday are substantiated there, and the electromagnetic theory of light was born.

The career of Maxwell had proceeded in the meantime. In 1856 he had left Cambridge to become professor of physics at Marischal College in Aberdeen, Scotland. He hoped the long vacation of Scottish universities would permit him to spend many months of each year at his beloved Glenlair with his father, who was in declining health. Unfortunately, his father died that same year.

In Aberdeen Maxwell met the daughter of the principal of his college, Katherine Mary Dewar, and married her. There are conflicting reports on the pleasantness of Maxwell's wife. She helped him in experiments on gas viscosity and color vision; his letters to her are affectionate and humorous, but every now and then, he goes into sermonlike reflections with ample quotations from the Bible, which seem to me exaggerated, even taking into account the time and place of his writing and Maxwell's deep piety. Mrs. Maxwell in later life was an invalid for extended periods, and her husband nursed her with great devotion, even at the time when he was mortally ill. They had no children.

As far as Maxwell's teaching is concerned, the weight of evidence shows that it was superior for pupils worthy of it, but not too intelligible for mediocre pupils, who made up the bulk of his classes. At Aberdeen he wrote, "I am at full college work again. A small class with a bad name for stupidity, so there was more field for exciting them to activity." Also, "No jokes of any kind are understood here. I have not made one for two months, and if I feel one coming I shall bite my tongue." For the whimsical Maxwell, this was a strong comment. Later, at Cambridge, he had pupils of better quality. Sir Horace Lamb (1849–1934), a distinguished mathematical physicist, was one of them, and he commented on Maxwell as follows:

> He had his full share of misfortune with the blackboard and one gathered the impression—which is confirmed I think by his writings—that though he had a firm grasp of essentials and could formulate great mathematical conceptions, he was not very expert in the details of minute calculations. His physical instinct saved him from really vital errors. . . . Maxwell's lectures had a great interest and charm for some of us, not so much for the sake of the subject matter, which was elementary, but in the illuminating glimpses we got of the lecturer's own way of looking at things, his constant recourse to fundamentals, and even his expedients when he was in difficulty, the humorous and unexpected digressions, the occasional satirical remarks and often a literary or even poetical allusion.

The Aberdeen appointment came to an end in an unusual way: Marischal College was merged in 1860 with King's College into one university, and only one professor for each subject was retained. In natural philosophy another man was preferred to Maxwell, who was dismissed. He had applied for a chair at Edinburgh, but there his friend Tait was preferred. However, shortly thereafter, Maxwell was appointed at King's College in London, where he remained until 1865. At that time, he retired again to Glenlair to write his celebrated treatise on electricity, one of the great texts in physics. During his Glenlair period he served as examiner or moderator at Cambridge in 1866, 1867, 1869, and 1870, instituting important and useful reforms in the tripos.

His second great paper in electricity appeared while Maxwell was in London, and it was followed in 1864 by "A Dynamical Theory of the Electromagnetic Field." In this paper the theory becomes much more abstract. The ether models are abandoned, and we find at the very beginning a clear statement of the aims of the paper and the equations of the field. In 1857 Faraday had written a letter to Maxwell that must have made a deep impression on the recipient. At its end, Faraday asked:

> There is one thing I would be glad to ask you. When a mathematician engaged in investigating physical actions and results has arrived at his conclusions, may they not be expressed in common language as fully, clearly, and definitely as in mathematical formulae? If so, would it not be a great boon to such as I to express them so? — translating them out of their hieroglyphics, that we also might work upon them by experiment. I think it must be so, because I have always found that you could convey to me a perfectly clear idea of your conclusions, which, though they may give me no full understanding of the steps of your process, give me the results neither above nor below the truth, and so clear in character that I can think and work from them. If this be possible, would it not be a good thing if mathematicians, working on these subjects, were to give us the results in this popular, useful, working state, as well as in that which is their own and proper to them?

Perhaps Maxwell had Faraday's letter in mind when, at the beginning of his paper, he explained what he wanted to do by describing the electromagnetic field. The following are his words:

> In order to bring these results within the power of symbolical calculations I then express them in the form of the General Equations of the Electromagnetic Field. These equations express —

(A)   The relation between electric displacement, true conduction, and total current, compounded of both.

(B)   The relation between the lines of magnetic force and the induction coefficients of a circuit, as already deduced from the laws of induction.

(C)   The relation between the strength of a current and its magnetic effects according to the electromagnetic system of measurement.

(D)   The value of the electromotive force in a body, as arising from the motion of the body in the field, the alteration of the field itself, and the variation of electric potential from one part of the field to another.

(E)   The relation between electric displacement, and the electromotive force which produces it.

(F)   The relation between an electric current, and the electromotive force which produces it.

(G)   The relation between the amount of free electricity at any point, and the electric displacements in the neighbourhood.

(H)   The relation between the increase or diminution of free electricity and the electric currents in the neighbourhood.

There are twenty of these equations in all, involving twenty variable quantities.

I then express in terms of these quantities the intrinsic energy of the Electromagnetic Field as depending partly on its magnetic and partly on its electric polarization at every point.

From this I determine the mechanical force acting, 1st, on a moveable conductor carrying an electric current; 2ndly, on a magnetic pole; 3rdly, on an electrified body.

For the modern reader who has a knowledge of electricity, the following are (in modern notation) the equations as set down by Maxwell in 1864:

(A) $\mathbf{j} = \mathbf{i} + \dfrac{\partial \mathbf{D}}{\partial t}$

(B) $\mathbf{B} = \text{curl } \mathbf{A}$

(C) $4\pi \mathbf{j} = \text{curl } \mathbf{H}$

(D) $\mathbf{E} = \mathbf{v} \times \mathbf{B} - \dot{\mathbf{A}} - \text{grad } \psi$

(E) $\mathbf{D} = k\mathbf{E}$

(F) $\mathbf{E} = -\rho \mathbf{i}$

(G) $e - \text{div } \mathbf{D} = 0$

(H) $\dfrac{de}{dt} + \text{div } \mathbf{i} = 0$

The reader will recognize here all the fundamentals of modern electricity.

The symbols used are $\mathbf{E}$, $\mathbf{D}$, $\mathbf{H}$, and $\mathbf{B}$ for the electric and magnetic force and induction. In an isotropic medium, $\mathbf{B} = \mu\mathbf{H}$ and $\mathbf{D} = k\mathbf{E}$. $\mathbf{A}$ is the vector potential. The current density due to conduction is $\mathbf{i}$, and the total current density is $\mathbf{j}$. The resistivity is ρ. Equation (D) contains the second circuital equation, usually written as curl $\mathbf{E} = -\partial \mathbf{B}/\partial t$ (as can be seen by combining equations (B) and (D)). The function ψ is the electrostatic potential, and e the electric charge. The units used are absolute electromagnetic units throughout, which hides the constant c. The same equations are found with some changes in the 1873 *Treatise*. It is in reference to them that Ludwig Boltzmann (1844–1906), quoting Goethe, said "War es ein Gott der diese Zeichen schrieb?" (Was it a God that wrote these signs?)

In London Maxwell also taught evening classes for working people and participated assiduously in the works of a committee including, among others, Balfour Stewart (1828–1887), Henry Charles Fleeming Jenkin (1833–1885), William Thomson, Ernst Werner von Siemens (1816–1892), and himself for the determination of electrical units, mainly the unit of resistance—the ohm. The measurement was to be an absolute one, followed by comparison of that unit to a certain mercury column. It was important work required by the beginning electrical industry and sponsored by several nations, at the prodding of major scientists and electrical engineers. William Thomson in England and Siemens in Germany were among them. Experience with the problems of electrical cables for telegraphy had amply demonstrated the urgency of the establishment of suitable measurement standards and methods. Ultimately, a standard ohm was adopted in 1881 by an international congress in Paris.

The question of units and dimensions was treated repeatedly by Maxwell, and one aspect of it had fundamental significance, as Maxwell clearly explains:

> There are two distinct and independent methods of measuring electrical quantities with reference to received standards of length, time, and mass.
>
> The electrostatic method is founded on the attractions and repulsions between electrified bodies separated by a fluid dielectric medium, such as air; and the electrical units are determined so that the repulsion between two small electrified bodies at a considerable distance may be represented numerically by the product of the quantities of electricity, divided by the square of the distance.
>
> The electromagnetic method is founded on the attractions and repulsions observed between conductors carrying electric currents, and separated by air; and the electrical units are determined so that if two equal straight conductors are placed parallel to each other, and at a very small distance compared with their length, the attraction between them may be represented numerically by the product of the currents multiplied by the sum of the lengths of the conductors, and divided by the distance between them.
>
> These two methods lead to two different units by which the quantity of electricity is to be measured. The ratio of the two units is an important physical quantity, which we propose to measure. Let us consider the relation of these units to those of space, time, and force (that of force being a function of space, time, and mass).
>
> In the electrostatic system we have a force equal to the product of two quantities of electricity divided by the square of the distance. The unit of electricity will therefore vary directly as the unit of length, and as the square root of the unit of force.
>
> In the electromagnetic system we have a force equal to the product of two currents multiplied by the ratio of two lines. The unit of current in this system therefore varies as the square root of the unit of force; and the unit of electrical quantity, which is that which is transmitted by the unit current in unit of time, varies as the unit of time and as the square root of the unit of force.
>
> The ratio of the electromagnetic unit to the electrostatic unit is therefore that of a certain distance to a certain time, or, in other words, this ratio is a *velocity;* and this velocity will be of the same absolute magnitude, whatever standards of length, time, and mass we adopt.
>
> The electromagnetic value of the resistance of a conductor is also a quantity of the nature of a velocity, and therefore we may express the ratio of the two electrical units in terms of the resistance of a known standard coil; and this expression will be independent of the magnitude of our standards of length, time, and mass.

Maxwell's electrical work was ultimately summed up in his *Treatise on Electricity and Magnetism*. The first edition appeared in 1873, the second in 1881, after Maxwell's death, and was edited by W. D. Niven, although it had been revised in part by Maxwell. The third edition, 1891, was edited by Joseph John Thomson (1856–1940). The book is not a systematic treatise in the ordinary sense, and it would be hard to go through it from one end to the other. It contains heaps of precious materials and insights, but the reader has to dig them out. Some chapters are of pure mathematical interest with little physics, while other chapters are devoted to detailed calculations of special problems or even to experimental details. Maxwell's equations appear only in Chapter IX of Part 4, well into the second half of the second volume. It is no wonder that the book appeared to contemporary physicists, and even to the immediate following generations, as a great but inaccessible monument. It is not an exaggeration to say that Maxwell's theory penetrated continental Europe through two great followers: Heinrich Rudolf Hertz (1857–1894) in Germany and Henri Poincaré in France (1854–1912). However, even today, I have consulted occasionally Maxwell's book for my own work and have found in it unexpected illuminations or help.

Maxwell had been at Glenlair since 1865, when in 1870 Cambridge University received a splendid gift from the seventh Duke of Devonshire, who had been a second wrangler in 1829, and was then Chancellor of the university. The duke, directly related to Henry Cavendish of electrical fame, gave the university 6,000 pounds sterling for the construction of a physics laboratory. Up until then there was none in Cambridge, and the various fellows experimented in their colleges. There was also to be a Cavendish professorship of experimental physics, with a salary of about £500 per year. William Thomson had been approached for the job, but he did not want to move from Glasgow; Helmholtz proved also unavailable; and, as a third choice, Maxwell, the greatest of the three, was elected in 1871. He thus initiated the series of Cavendish professors, without doubt the most illustrious

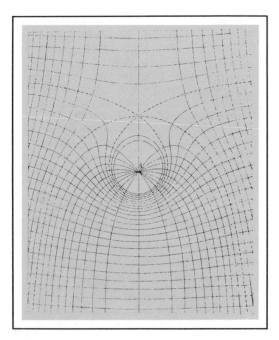

One of the beautiful drawings in Maxwell's Treatise on Electricity and Magnetism *(1873), showing the magnetic field produced by a wire and a uniform field.*

The facade of the Cavendish Laboratory in Cambridge, as it appeared from Maxwell's time onward. The building is devoted to other university uses now, and the Cavendish Laboratory has been relocated. (Cavendish Laboratory, Cambridge University.)

sequence in modern physics. On Maxwell's death in 1879, John William Strutt, Lord Rayleigh (1842–1919) got the job, which he resigned in 1884, to be followed by J. J. Thomson. When he retired in 1919, Ernest Rutherford (1871–1937) followed. On his death, Sir William Lawrence Bragg (1890–1971) occupied the chair, to be followed in 1952 by Nevill Francis Mott.

Maxwell devoted great energy to the building and equipment of the new laboratory. It was ready for occupation and functioning by 1874. He had started lecturing there in October 1871, and in his long inaugural lecture, he expresses in detail his philosophy of teaching physics and the function of the laboratory as a demonstration and research institution.

It seems that the announcement of the inaugural lecture was fouled up, possibly with Maxwell's connivance, so that very few attended, while the bigwigs of the university attended the second lecture to be taught the elements of thermometry. Maxwell liked such jokes. In the inaugural lecture, however, one finds some remarkable utterances on various scientific subjects. As a sample I report the following remark that may sound prophetic of things to come in quantum mechanics:

> The theory of atoms and void leads us to attach more importance to the doctrines of integral numbers and definite proportions; but, in applying dynamical principles to the motion of immense numbers of atoms, the limitation of our faculties forces us to abandon the attempt to express the exact history of each atom, and to be content with estimating the average condition of a group of atoms large enough to be visible. This

method of dealing with groups of atoms, which I may call the statistical method, and which in the present state of our knowledge is the only available method of studying the properties of real bodies, involves an abandonment of strict dynamical principles, and an adoption of the mathematical methods belonging to the theory of probability. It is probable that important results will be obtained by the application of this method, which is as yet little known and is not familiar to our minds. If the actual history of Science had been different, and if the scientific doctrines most familiar to us had been those which must be expressed in this way, it is possible that we might have considered the existence of a certain kind of contingency a self-evident truth, and treated the doctrine of philosophical necessity as a mere sophism.

During his last Cambridge period, Maxwell also undertook the publication of Cavendish's electrical papers. The manuscripts had been lying dormant for about a century; they contained first-class work that undoubtedly would have influenced the development of electrical science had it been known. Cavendish's idiosyncrasies kept it from being published.

In the summer of 1879, when he was forty-eight years old, Maxwell showed alarming symptoms of the same abdominal disease that had killed his mother when he was eight. It was apparent that he had little time to live, and he expired on November 5, 1879, the year of Einstein's birth.

Let us now consider Maxwell's other great achievement — statistical mechanics. The study of Saturn's rings had prepared his mind for statistical problems. In 1859 the British Association met in Aberdeen, his territory, where he had to give an address. He chose as his subject "Illustrations of the Dynamical Theory of Gases," which was not a new subject, as we shall see in another chapter. For example, there was a dynamical explanation of the gas laws that went back about one hundred years to Daniel Bernoulli (1700–1792). However, nobody had tackled the problem of the velocity distribution among the molecules of a gas. In two pages, Maxwell solves this fundamental problem. It is true that his arguments leave much to be desired, but their simplicity is astonishing and the result is correct. (Maxwell's words are given in Appendix 10.) The figure on page 169 shows the number of molecules between velocity v and $v + dv$. It is remarkable that this figure is universal, in the sense that it is valid for all gases and all temperatures, provided the scales for the abscissas and ordinates are suitably chosen. The result has striking similarity with the "error curve," according to which errors are distributed among observations in the method of least squares. The same law gives the distribution of hits on a target by a gun pointed in a constant direction at the center of the distribution, that is, the so-called "dispersion pattern."

We can trace the source of Maxwell's interest in kinetic theory to three sources: his studies on Saturn's rings, already mentioned; the reading of early papers by Rudolf Clausius (1822–1888) on the subject; and readings of Laplace and George Boole's (1815–1864) books on probability theory, as well as of a long essay by Sir John Herschel (1792–1871), reviewing a book by the statistician Lambert-Adolphe-Jacques Quetelet (1796–1874).

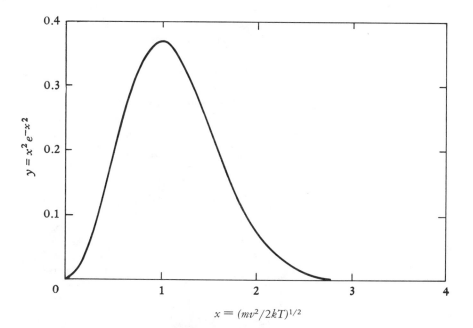

$$x = (mv^2/2kT)^{1/2}$$

The Maxwellian distributions of molecular velocities in a gas. The curve of equation $y = x^2 exp(-x^2)$ can be used for any molecular mass m and absolute temperature T, by making $x = (m/2kT)^{1/2}v$. Here k is the Boltzmann constant: $k = 1.38 \times 10^{-16}$ erg/K. Out of N molecules, the number dN of those having a velocity between v and $v + dv$ is $dN = N(8m/\pi kT)^{1/2}y\ dv = (N/\sqrt{\pi})4y\ dx$. (From From X-Rays to Quarks *by Emilio Segrè. W. H. Freeman and Company. Copyright © 1980.)*

In 1857 Clausius had given a derivation of the gas laws based on a kinetic model, in which he also considered rotational and possible internal motions of the molecules, improving on earlier work that used simpler point molecules as models. The paper, entitled "The Kind of Motion We Call Heat," was widely read and translated into English. Soon thereafter, he introduced the concept of mean free path, but he, as did every other writer on the subject, used only mean values of the velocities or similar magnitudes. Through the efforts of Clausius and others, some well-defined problems could be attacked, in the hope of obtaining results comparable with experiment. Generally, the qualitative elementary theory was simple, but more precise results required difficult and laborious calculations in which the velocity distribution was of paramount importance.

For instance, the viscosity coefficient μ of a gas is easily written once the concept of mean free path is introduced, as $\mu = (1/3)\rho l \langle v \rangle$, where ρ is the density of the gas, l the mean free path, and $\langle v \rangle$ the average velocity.

Maxwell derived that equation together with other "transport" equations. The fundamental idea is that a molecule starting in one layer transports to another layer the quantity (momentum, energy, etc.) characteristic of the first layer, but that this transport is limited to a distance equal to the mean free path. In the formula for the viscosity, the quantity transported is the momentum in a certain direction. Since l is inversely proportional to the density, the viscosity is independent of pressure and should depend on the temperature through v and, hence, be proportional to $\sqrt{T}$. Old measurements by Thomas Graham (1805–1869) on gaseous diffusion and by Stokes on gas viscosity permitted an estimate of l for a gas at atmospheric pressure; 6×10^{-6} cm is a representative number, calculated by Maxwell in his paper of 1859.

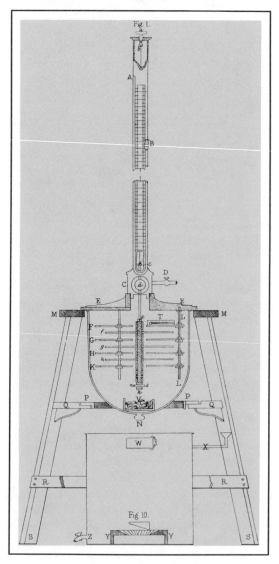

The apparatus with which Maxwell and his wife measured the temperature and pressure dependence of the viscosity of air. The primary observation was the damping of the oscillations of a torsion pendulum, produced by air friction. The apparatus is now exhibited at the Cavendish Laboratory. (From Philosophical Transactions of the Royal Society, *vol. 156. Courtesy of the University of California, Berkeley.)*

Maxwell with his wife and his dog, Toby (Courtesy of C. W. F. Everitt.)

Maxwell wrote to Stokes in 1859, commenting, "This is certainly very unexpected, that the friction should be as great in a rare as in a dense gas. The reason is, that in the rare gas the mean path is greater, so that the frictional action extends to greater distances." Maxwell tried to verify his theoretical results by experiment, measuring the damping of a complex of discs. He was helped in these experiments by his wife. When it came to checking the temperature dependence, they kept the experimental room as cold as an English winter (they were in London) would permit, and then heated the room to as high a temperature as they could obtain and/or withstand. The independence of the viscosity from the pressure was well verified, but the dependence on temperature proved to be linear and not proportional to the square root. Much later, Maxwell found the recondite explanation for this fact. Those experiments made it possible to measure the mean free path from the value of μ and the velocity as obtained from the velocity distribution law. The mean free path came out to be 5.6×10^{-6} cm. That result is very important because it provides a handle for finding the size of a molecule, its mass and Avogadro's number. All those quantities were not known even as an order of magnitude, and it signified great progress to know something about them. Diffusion experiments confirmed the results obtained by viscosity experiments, which strengthened the confidence in the results.

A typical postcard from Maxwell, signed dp/dt. *This is the equivalent of JCM according to Clapeyron's equation (in the notation used by Maxwell). (Courtesy of C. W. F. Everitt.)*

By 1873, at the meeting of the British Association in Bradford, Maxwell could report fair numerical values for molecular magnitudes. They were due to the pioneer work of Johann Joseph Loschmidt (1821–1895), followed by Maxwell and others. They are reproduced in the following table:

| | H_2 | O_2 | Present | |
			H_2	O_2
$\langle v^2 \rangle^{1/2}$	1859 m/sec	465 m/sec	1839 m/sec	461 m/sec
mean free path	9.65×10^{-6} cm	5.6×10^{-6} cm	11.2×10^{-6} cm	6.6×10^{-6} cm
collisions/sec	17.7×10^{9}	7.64×10^{9}	16.6×10^{9}	7.0×10^{9}
diameter	5.8×10^{-8} cm	7.6×10^{-8} cm	2.72×10^{-8} cm	3.75×10^{-8} cm
mass	4.6×10^{-24} g	—	3.34×10^{-24} g	26.57×10^{-24} g

Present values are within about 10 percent of Maxwell's numbers although where the magnitudes are well definable, we now know the numbers with much more precision.

Another topic of even more fundamental importance covered in the 1860 papers was what is now known as equipartition of energy. Maxwell showed by direct calculation that the kinetic energy associated with each degree of freedom for a point molecule is independent of its mass. Also, for more complicated molecules, each degree of freedom showed on calculation the same kinetic energy. This result of the calculations conflicted with experiment and presented a most serious difficulty. The final explanation had to await quantum theory.

Maxwell returned to the kinetic theory in 1865, when he gave the Bakerian lecture on his viscosity experiments and in a great paper of 1866 in which he recast the whole theory on a sounder basis. That paper was compared by Boltzmann, who in the meantime had entered the field, to a great musical masterpiece. One of the features of the paper is that using a repulsion law with a force proportional to the inverse fifth power of distance, the calculations are simplified in a surprising way. That repulsion law is frequently called Maxwellian repulsion; it often permits calculations of general significance by first performing the calculation for the special

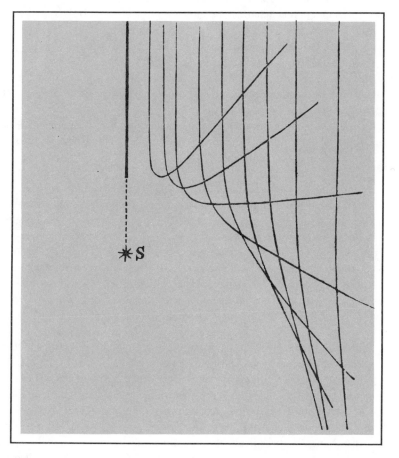

Trajectories of a particle repelled inversely to the fifth power of its distance from a fixed center. This special repulsion law substantially simplifies calculations in kinetic theory, but is not essential for the final statistical result. Boltzmann was deeply impressed with Maxwell's cleverness in devising this method. (Illustration from "On the Dynamical Theory of Gases," Philosophical Transactions, *vol. 157.)*

case, and later proving that the specific repulsion law is unimportant. The later Maxwell papers on kinetic theory are difficult reading, but they touch on deep questions, some of which are still unsolved.

In 1875 Sir William Crookes (1832 – 1919), the renowned British chemist and physicist, invented the radiometer, a little instrument still sold in hobby shops. It is composed of an evacuated glass bulb containing a light rotatable axle to which vanes, black on one side and reflecting on the other, are attached. The axle and vanes, when exposed to light, turn rapidly. At first it was thought that electromagnetic radiation pressure, a prediction of Maxwell's electromagnetic theory, was the cause of the motion. That proved to be incorrect. The effect is due to the presence of a rarefied gas in the bulb. Its explanation is subtle and full of pitfalls. Maxwell tackled it, and, in so doing, founded the dynamics of rarefied gases. That subject branched into technical applications that go from astronautics to isotope separation. It is a tribute to Maxwell's genius that almost a hundred years after their composition, his papers were still studied as living physics and inspired new researches.

Electromagnetism and statistical mechanics were the two major endeavors of Maxwell, but he also made important discoveries in many other physics fields, where a number of theorems or formulas carry his name. He left the mark of his genius in such diverse subjects as thermodynamics, geometrical optics, elasticity, cybernetics (on that subject he wrote a truly fundamental paper), engineering, statics, physiological and geometrical optics, and others.

Maxwell's electromagnetic theory did not easily penetrate contemporary physics. It appeared mathematically difficult, and it had relatively minor experimental support; it had as rivals many other theories of electricity based on better-known Newtonian ideas, and, thus, especially on the continent, it was resisted. In England some engineers, such as Oliver Heaviside (1850 – 1925), plumbed it deeply, but even Lord Kelvin, as late as 1904, when his Baltimore lectures of 1884 were published, was still skeptical about the electromagnetic theory of light.

In 1870, when the magazine *Nature,* which is still flourishing today, was founded, Maxwell became one of its frequent contributors, and the *Encyclopaedia Britannica* also contained articles by him, some true masterpieces of the expository art.

Although Maxwell enjoyed a distinguished reputation during his lifetime, I doubt that he was fully appreciated by his contemporaries. Compared to his friend William Thomson, certainly a great mathematical physicist, we find Thomson, a professor at Glasgow at twenty-two, a position which he enjoyed all his long life, and a knight at forty-two. Possibly this was due to the great practical applications of Thomson's work and to his personality. Maxwell was more withdrawn, occasionally biting, and occupied with difficult and fundamental theories, rather than with a succession of brilliant problems. It is, however, remarkable that Maxwell himself considered Thomson at least his equal. In his youth, the age difference and the precocity of Thomson can explain why Maxwell considered him a sort of older brother and turned to him for advice, but posterity has recognized the greater profundity of Maxwell. Helmholtz too, also a great physicist, enjoyed higher

*Heinrich Hertz (1857–1894).
(Photograph courtesy of Deutsches
Museum, Munich.)*

reputation than Maxwell. His universality puts him in a category by himself, but his imprint on future physics is not comparable to that of Maxwell.

If Maxwell had lived as long as Lord Kelvin, he would have seen Hertz's discovery of electromagnetic waves and even their application to transatlantic communication (1902), relativity, and the discovery of the quantum.

Possibly the man who did most to secure the triumph of Maxwell's electromagnetism was Heinrich Hertz. He was born in a well-to-do family of the high bourgeoisie at Hamburg. His father was of Jewish origin, but the family had been baptized and tended to complete assimilation. The father was first a lawyer, then a judge, and ultimately one of the Hamburg Senators, a member of the governing body of the Hanseatic city. We find among the German physicists, mathematicians and professors in general, a number of prominent Jews. The pattern is frequently that of an emancipated, successful, and rich father followed by children eminent in intellectual professions. In Imperial Germany, the army, the highest social echelon, was practically precluded to Jews, and the socially desirable position of a university professor (they became Geheimrath and Exzellenz) was a way to elevated social status. In fact, it is pathetic to see great men decorate themselves with court titles, when, in fact, they should have been the honor of the title and not vice versa, but even great men had their weaknesses.

The refined, active, and noble way of life of the Hertz family in the second half of the nineteenth century can be followed in memoirs, letters, and diaries that were published in 1927. The Hertz family, perhaps conscious of its own importance, preserved letters and diaries. Thus, we can trace even the early youth of Heinrich, who was a sort of prodigy, both intellectually and in manual ability. He was always the first in his class at highly competitive schools, learned many languages, including Arabic, would go on Alpine tours with his father, and recite Homer in Greek and Dante in Italian. In spite of all this, he was not a pedantic boy, and he certainly had a sense of humor in his letters. At the university he tried to study engineering, but it would have been easy to predict that he would turn to science. The letter in which he asked his father permission to change from engineering to physics (he was twenty years old) is an eloquent statement of a problem that has confronted many scientists at the beginning of their careers, when practical considerations pull one way and desire the other, and even a Hertz could not be sure of his own ability.

We also see the usual pride of a young man suffering from being financially dependent upon his good father (who had plenty of money anyway), and was more than happy to subsidize the brilliant son. In 1878 Hertz transferred from Munich, where he had begun his studies, to Berlin. The great Helmholtz, who was a professor there, spotted him immediately and gave him a subject for experimental research in electricity (to find, if possible, an inertial mass of the electricity as manifested in the phenomena occurring at the closing or opening of a circuit). By 1880 Hertz was already Helmholtz's assistant, and in the letters of the younger man, we have a glimpse of Helmholtz as a person, of his house, of his lectures and his science, while in letters from Helmholtz, Hertz progresses from "Geehrter Herr Doktor" to "Verehrter Freund" to "Geehrter Freund und Kollege." Hertz's scientific work is based on the combination of unusual analytical power with extraordinary experi-

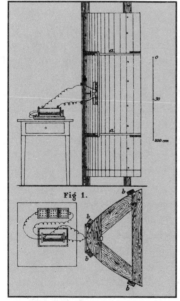

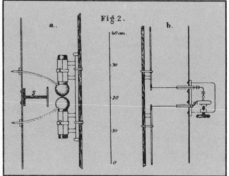

The instruments used by Hertz for demonstrating electromagnetic waves. The spark produces strongly damped waves of some decimeters wavelength in Fig. 1 *and* Fig. 2a *(note the parabolic reflector). The detector is a spark in a resonating circuit* (Fig. 2b). *Observing the peculiarities of the spark, Hertz was led to the discovery of the photoelectric effect. The drawing is by Hertz from the* Annalen der Physik und Chemie *(vol. 36, 1889).*

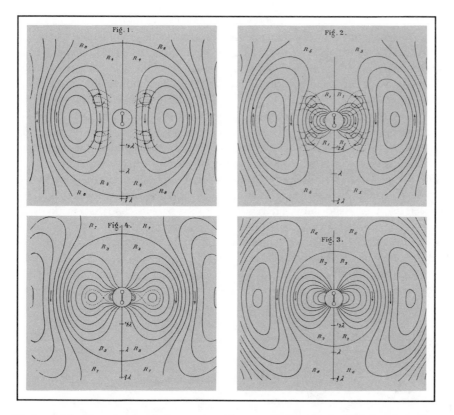

The lines of electric and magnetic force produced by an oscillating electric dipole in a region near the source. Note the wavelength of the emitted radiation marked on the figures. The figures represent the field at times 0, $\lambda/4c$, $\lambda/2c$, $3\lambda/4c$. The drawings are by Hertz from the Annalen der Physik und Chemie *(vol. 36, 1889).*

mental ability. At first he treated classical problems in elasticity, obtaining original results that have even practical importance, and he performed experiments on gas discharges, where he reached wrong conclusions on the nature of cathode rays. His most important work started in 1886, when he became professor at Karlsruhe at the Polytechnical School. He had previously published theoretical papers, trying to find proofs for Maxwell's theory and trying to connect them to the electrodynamical theories of Wilhelm Weber and Carl Neumann. In these studies the necessary existence of electromagnetic waves loomed large. It also became apparent that their frequency was of paramount importance in any experimental investigation, and, thus, Hertz had to find a way of obtaining very high frequency oscillations. He succeeded in October of 1886, seven years after Maxwell's death.

The oscillating circuit was what we would call today a dipole fed by a spark, and the detector was a circular piece of wire with a small gap in which a spark appeared when it was excited by an electromagnetic wave. With these tools Hertz could reproduce many of the fundamental experiments of optics and demonstrate that his waves propagate with the velocity of light. He considered his experiments conclusive proofs of Maxwell's theory. On the other hand, he does not seem to have anticipated their enormous practical importance for communications. It was only during the year of Hertz's death that the young Guglielmo Marconi (1874–1937), inspired by readings of Hertz's work, and perhaps by conversations with Professor Augusto Righi (1850–1920), who had extended Hertz's investigations on electromagnetic waves, started a conscious effort to apply Hertzian waves to telegraphy.

Hertz published one paper after the other, and often communicated his results to his mentor, Helmholtz, before publication. Through the formal German of his letters, and the answers of the Geheimrat, one easily recognizes the warm rapport between the master and pupil. Hertz's letters to the parents exude satisfaction and pride and again show a close and affectionate relationship.

Besides the discoveries concerning electromagnetic waves, Hertz soon observed a phenomenon of an entirely different sort. Light, and especially ultraviolet light, facilitated the sparks in the gap of his receivers. Hertz was puzzled by the observation, and he described it to his father. When the latter asked about its importance, Heinrich mentioned Faraday's discovery of the relation between magne-

A postcard dated November 7, 1887, from Helmholtz to his former pupil Hertz, referring to a paper on electromagnetic waves and stating, "Bravo!! Will forward it to the printer on Thursday." (Photo by Deutsches Museum, Munich.)

A page of the diary kept by Hertz, March 19, 1888. (Photo by Deutsches Museum, Munich.)

tism and light. Hertz could not have imagined that he had seen a manifestation of what is now called the photoelectric effect, one of the links between classical and quantum physics and one of the observational facts that were to show the inadequacy of the electromagnetic theory of light. Furthermore, in 1892 Hertz's assistant, Philipp Lenard (1862–1947), who in his old age became an ardent Nazi and besmirched his earlier brilliant career with scientific nonsense and human depravity, discovered another important fact. He succeeded in making an extremely thin aluminum window on a discharge tube; the window was so thin that electrons (which had not yet been discovered) could pass through it and be observed in air. Hertz immediately perceived the importance of the experiment and reported it to Helmholtz. Hertz's experiments on the optical properties of electromagnetic waves became very fashionable. He had refracted his waves with a huge pitch prism, reflected them on conducting mirrors, and polarized them by passing them through gratings of parallel wires. Those experiments were repeated and supplemented by others in many laboratories.

Soon Hertz's fame extended worldwide and the ever-vigilant Geheimrat Friedrich Theodor Althoff decided to promote him to a better place than Karlsruhe and to foster his research as much as possible. Althoff is a very interesting figure in his own right. He was the most powerful bureaucrat in the Prussian Ministry of Education, being in charge of universities. He was extremely intelligent, shrewd, totally devoted to his self-assigned task to make the German universities the best in the world, and a complete autocrat. He had his plans and seemed to use the professors as pawns in an intricate chess play. At the same time, he had uncanny judgment and benevolence for his wards. Hertz's letters to his parents describing his encounters with Althoff give an insight into German universities at their best, as

The house in Bonn inhabited in succession by Clausius and Hertz. (Photo by Deutsches Museum, Munich.)

vivid and faithful as any. Althoff transferred Hertz to Bonn as a successor to Clausius, whose house Hertz occupied. The move delayed the beginning of experimental work, and Hertz turned to theory. There was a demand for the publication of a book containing his memoirs on electricity, a rare honor repeated in 1928 when it became necessary to republish the collected memoirs of Erwin Schroedinger (1887 – 1961) on quantum mechanics, and Hertz published a comprehensive paper on electrodynamics, entitled *Untersuchungen über die Ausebreitung der elektrischen Kraft,* which became the standard work on the subject for many people who could not fathom the original Maxwell. Hertz dedicated the collection of his works to Helmholtz, whose inspiration he felt deeply. Later, he tackled another fundamental problem — the electrodynamics of moving bodies. He was faced there with difficulties that could not be surmounted at the time; it was only fifteen years later that Einstein found the key. Hertz also tried a reformulation of classical mechanics. The concept of force, from the time of Newton on, had given rise to epistemological problems. Hertz tried to eliminate them by considering the action of one system on another directly. The attempt is elegant, but it has not affected the further development of mechanics, which has turned in completely different directions with relativity and quantum mechanics.

When everything seemed propitious to Hertz and his family, tragedy struck. An insidious disease — a bone malignancy — attacked his skull. From about 1892 on, he suffered excruciating physical pain and an equally tragic alternative of hope and despair, which he bore with great fortitude until his death on January 1, 1894, at the age of thirty-six. Helmholtz, who survived him by only a few months, commented, "In classical times one would have said that he was sacrificed to the envy of the gods."

Max Planck (1858–1947), who was only one year younger than Hertz, read his eulogy at the Physical Society of Berlin; his words forcefully conveyed his deep emotion. The thorough appreciation of Hertz's work and personality by such a contemporary is the best testimonial to his position among the physicists of his generation.

A Bridge to Modern Physics: H. A. Lorentz

The next step in electricity was accomplished by Hendrik Antoon Lorentz. At the beginning of this volume we found the great Dutch physicist Huygens as a transition figure between the epochs of Galileo and Newton. I end this chapter with another great Dutch physicist, also a transition figure, between classical physics and the "modern" physics of relativity and quantum mechanics. His position was well delineated by Einstein, who belongs to the generation following him, since he was twenty-six years younger than Lorentz. Einstein says:

> At the turn of the century, H. A. Lorentz was regarded by theoretical physicists of all nations as the leading spirit; and this with the fullest justification. No longer, however, do physicists of the younger generation fully realise, as a rule, the determinant part which H. A. Lorentz played in the formation of the basic principles of theoretical physics. The reason for this curious fact is that they have absorbed Lorentz' fundamental ideas so completely that they are hardly able to realise to the full the boldness of these ideas, and the simplification which they brought into the foundations of the science of physics.

Lorentz's roots are in Fresnel and Maxwell, with his uppermost branches reaching toward Planck and Einstein.

Lorentz was born in 1853 and died in 1928. I saw him, although I did not talk to him, presiding at the International Physics Conference at Como in 1927 in commemoration of the one hundredth anniversary of Volta's death. Some of my friends have had personal contacts with him, although only when he was relatively old, and they spoke of Lorentz with awe. He was a national glory in a country leery of hero worship, because he epitomized the best of the country: intellectual brilliancy, love for peace, democratic spirit, calm and cool-headed serenity, and shrewdness in human affairs.

Born at Arnhem, a provincial town in eastern Holland where he spent his youth, Lorentz was the son of a nurseryman. As soon as he went to school, he was recognized as the first of his class, and without apparent effort, he remained an exceptional pupil during his entire scholastic career. In 1870 he went to study at the University of Leyden, and in 1871 he passed *summa cum laude* his exam to become a candidate in mathematics and physics. His reputation had spread, but the mathematics examiner, although satisfied, said that his high expectations had been slightly disappointed. He discovered, however, that he had used questions not for the candidate exam, but for the doctor's degree.

Lorentz remained at Leyden only a short time. He returned to Arnhem and prepared his thesis there, while teaching at the local high school. He was, thus, to a large extent, self-taught. His guides were Maxwell's papers of which he commented "It was not always easy to understand Maxwell's thoughts," and Fresnel's works. Following Dutch custom, he defended publicly his dissertation "On the Reflection and Refraction of Light," and in December 1875 he obtained his doctorate from the University of Leyden. In his thesis we find the origins of the reformulation of electricity known as the "theory of electrons." I cannot describe that work better than Einstein did, who studied it carefully in his youth. His words were,

When H. A. Lorentz started to write, Maxwell's theory of electromagnetism was already generally known. This theory, however, suffered from a curious fundamental complexity, which prevented its essential features from being conspicuously clear. True, the vectorial field concept had ousted that of action at a distance; but the electric and the magnetic field were not yet conceived as being original entities, but as conditions governing ponderable matter, which was treated as existing in a continuous form. As a result, the electric field appeared to be split into the vector of the electric field strength and the vector of dielectric displacement. In the simplest case, these two elements were linked together by the dielectric constant; in principle, however, they were regarded and treated as two independent entities. Similar ideas prevailed in respect of the magnetic field. In conformity with this basic conception, empty space was treated as being a special case of ponderable matter, in which the relation between field strength and displacement only appeared to be particularly simple. And, more especially, this conception entailed that the electric and the magnetic field could never be conceived as independent of the state of movement of matter, which was regarded as the carrier of the field.

A study of Heinrich Hertz's investigation into the electrodynamics of moving bodies will give the reader a clear insight into the conception, prevalent at that time, concerning the electrodynamics of Maxwell.

It was here that H. A. Lorentz's act of intellectual liberation set in. With great logic and consistency he based his investigations on the following hypotheses: The seat of the electromagnetic field is empty space. In this field there is only *one* electric and *one* magnetic field vector. The electromagnetic field is created by atomistic electric charges, to which the field, in turn, reacts pondero-motorically. The only link between the electro-motoric field and the ponderable matter lies in the fact that elementary electric charges are intimately bound up with the atomistic constituents of matter. For the latter, Newton's law of motion is valid.

On this—thus simplified—foundation Lorentz based a complete theory covering all electromagnetic phenomena known at the time, including the electrodynamics of

Monument to H. A. Lorentz in the city of Arnhem. (Courtesy Gemeentearchief Arnhem.)

moving bodies. It is a work of rare clarity, logical consistency, and beauty, such as has been achieved only rarely in any empirically based science. The only phenomenon whose explanation could not be given completely—i.e. without additional assumptions—was the famous Michelson–Morley experiment. But it would have been unthinkable for this experiment to lead to the special relativity theory without the localisation of the electromagnetic field in empty space. The really essential step forward, indeed, was precisely Lorentz' having reduced the facts to Maxwell's equations concerning empty space, or—as it was then called—the ether.

That page, written by somebody who clearly knew his electricity, should be pondered by teachers of electromagnetism and by those who dabble in electric units and their dimensions.

Lorentz's theory of electrons was developed over many years. After his thesis, a great memoir of 1892 extended it and applied it with brilliant success to many optical phenomena; later, in 1895 and 1904, he introduced the famous Lorentz transformation, harbinger of relativity. The coronation and, at the same time, the limit of the theory was reached with the explanation of the Zeeman effect, discovered in 1896 in Leyden. While this discovery gave precious information on the carriers of the charges, what we call electrons, it also showed features that transcend the classical explanation.

In 1878 Lorentz was appointed professor of theoretical physics (the first such chair in Holland) at Leyden, and he remained there for the next thirty-four years. His life was very methodical and passed between three cities: Arnhem, Leyden, and, later, Haarlem, all in a radius of fifty miles. Life in Holland was provincial; everybody knew everybody else, and with the distances between centers being so small, there was a great deal of visiting between families and friends. On the other hand, the small number of people using the Dutch language forced the learning of foreign languages. Lorentz was exceptionally proficient in this and became fluent in French, English, and German, which was to become very important for him later. Lorentz married in 1881 a relative of one of his professors and had several children and a happy family life. Among his friends or contemporaries were Johannes Diderik van der Waals (1837–1923), who became professor at Amsterdam, having declined the Leyden chair that Lorentz took, and Kamerlingh-Onnes, who also obtained a chair at Leyden.

In his teaching Lorentz distinguished himself by the clarity of his lectures and by the interest of the subjects he treated, often at the forefront of research. He also published a series of texts on physics and mathematics to help beginning students. On his personal relations with students, there are conflicting tales. There is no doubt that he helped many and that he had affectionate and admiring pupils, but, equally certain, for all his courtesy, he was very reserved and almost aloof. I heard from various sources stories that more or less portray an enthusiastic student, thinking of having found something interesting and possibly new, going to Lorentz, who would listen and then pull out of a drawer a calculation, saying "I think I had the same result years ago; it must be correct." He was benevolent, but such events without much human warmth were not encouraging for a student.

Lorentz worked indefatigably for many years in the quiet of Dutch life. He communicated with his contemporaries only through their published works, but he does not seem to have made any effort to meet them personally. He could have encountered Maxwell, Helmholtz, or Hertz, but he did not meet any of them. He was happy in his privacy at Leyden, where he could devote all his time to a variety of physical subjects. Besides the theory of electrons, he studied thermodynamics and statistical mechanics, finding new results and using methods of his own invention, of notable clarity, rigor, and subtlety. In electrodynamics he developed the famous Lorentz transformation (1895), which was to prove fundamental for relativity.

His reputation continued to spread, and he received invitations to attend conferences, first in Germany and then in France. In 1902 he shared with Zeeman the Nobel Prize. Other aspects of Lorentz's personality and abilities came to the forefront. The Dutch physicist seemed to be a trilingual physics encyclopedia of unparalleled completeness and clarity. He also had tact, courtesy, and rapidity in understanding. He thus became the natural choice for presiding over international conferences. In that capacity, he did not limit himself to organizational matters or to presiding at the meetings, but contributed vitally in preparing the scientific programs, choosing the speakers, and summarizing the results. That activity, carried on in his superior manner, became an important occupation. The highlights were the Solvay Councils, the first of which was held in 1911. The industrialist Ernest Solvay (1838–1922) had donated a sizable sum for the gathering every two years of an assembly of about thirty leading physicists for the discussion of a physics theme of current importance. The first meeting was devoted to the quanta, and Lorentz presided at all conferences until 1927.

By now we are beyond the limits of this volume and we have indicated the passage from classical to modern physics.

Planck's constant h entered the scene in 1900, and in 1905 the epochal papers of Einstein signaled the passage. Lorentz obviously had to study Einstein; after all, one of the greatest of Einstein's discoveries was to give an unexpected interpretation to Lorentz's transformation and also to kill the ether, a mainstay of the theory of electrons. Lorentz and Einstein became close friends. The latter traveled many times to Leyden, and when Lorentz retired from Leiden, he tried hard to induce Einstein to become his successor. Einstein spoke of him in terms of great affection, and occasionally he signaled the importance of Lorentz's conceptions by calling the fundamental equations of electricity the "Maxwell-Lorentz equations."

In 1912 Lorentz left his Leyden professorship and took positions in Haarlem as secretary of the Hollandsche Maatschappij van Wetenschappen (Dutch Company of Sciences) and curator of Teyler fysisch Kabinet. The institutions were old ones and resembled the Royal Institution of Great Britain, where Faraday had labored so fruitfully. Lorentz was relieved of most official and administrative functions to be able to devote his time to research or whatever he liked. However, he kept a tie to Leyden by teaching there every Monday in a physics seminar that soon became world famous. The lectures on current subjects were attended by many visitors and were followed by lively discussions.

Lorentz traveled regularly to the United States, the first time in 1906 to Columbia University in New York, the last time to the California Institute of Technology in Pasadena in 1927. He was an important influence in bringing modern physics to the United States.

Another activity of a very practical type and of considerable importance that occupied Lorentz from 1920 to 1926. The Dutch, in their eternal struggle against the sea, decided to build a major dam across the Zuider Zee, in order to reclaim an important area of land and protect against inroads of the North Sea during gales. The planning of this major hydraulic work required a prediction and evaluation of its effects on the tides. Lorentz became the moving spirit of that theoretical engineering study with notable practical success. I wonder how much simpler the work would have been had he had modern computers.

In his main work, the electron theory, Lorentz, like Moses, arrived in sight of the promised land but could not enter it. Classical physics could not surmount the ultimate difficulties presented by the explanation of the spectra. The door was opened only in 1913, when Niels Bohr (1885 – 1962) combined Planck's quantization ideas with the atomic model suggested by Rutherford on the strength of the latter's studies in radioactivity. Lorentz was able to see the seeds planted by his work blossom into the vigorous plants of relativity and quantum mechanics, but of this he was mainly a spectator.

Chapter 5

Heat: Substance, Vibration, and Motion

Between Chemistry and Physics:
The Properties of Gases

Heat phenomena have been known since the beginning of civilization. Any technology of metals requires application of heat, and empirical notions on the subject must have accumulated very early. The concept of temperature is rooted in our immediate perceptions of hot and cold. Furthermore, any chemical operation involves heat generation or absorption. The most conspicuous effect is "fire" resulting from oxidation. Thus, the study of heat is intimately connected with chemistry. Moreover, for a long period, the scientific study of heat was intertwined with the study of gases, the center of attention in the eighteenth century. We could arbitrarily date the initiation of the scientific study of heat from the construction of the first thermometers by the Accademia del Cimento, around 1650. Earlier Galileo had built a thermoscope, but the members of the accademia, mostly his disciples, made an extensive and systematic use of the thermometer. The instrument, however, had no fixed points, and different instruments could not be directly compared.

Newton in 1701 proposed a scale in which the freezing point of water was taken as zero and the temperature of the human body as twelve. The expansion coefficient of the fluid used in the thermometer was, of course, assumed to be constant, or, rather, the temperature was defined in such a way as to make the expansion coefficient constant. Daniel Gabriel Fahrenheit (1686–1736), a little later, proposed to assume as zero the lowest temperature then attainable, by a mixture of ice and salt. Shortly after Fahrenheit's death, the fixed points were chosen as the freezing temperature of water and the temperature of boiling water at standard atmospheric pressure (the Celsius scale). For a more scientific scale, independent of any specific substance, one had to wait about a hundred years, as we shall see.

The concept of quantity of heat, separate from that of temperature, also grew in the eighteenth century. Joseph Black (1728–1799) did much to establish it.

He was of Scottish-Irish origin, educated in Ireland and at Glasgow, the son of a prominent wine merchant at Bordeaux. He became a lecturer in chemistry at Glasgow in 1756, where he later taught anatomy, medicine, and, ultimately, chemistry. At that time one could be an expert on *all* these subjects. He wrote only three papers, but he became famous for his lectures and an extensive correspondence. His permanent contributions are mostly in chemistry, but in physics, he clarified the concepts of heat capacity, specific heat, and latent heat of melting, which he measured for water, although not accurately. His theoretical ideas were still dominated by the concept of phlogiston.

The doctrine of phlogiston (from the Greek φλόγιστον, meaning combustible) originated with Johann Joachim Becher (1635–1682), a German who had studied medicine, but worked in many fields and was on the fence between alchemy and chemistry. The doctrine was developed to a high degree by George Ernst Stahl (*ca.* 1660–1734), also a German medical man and chemist. In that theory there was a "substance," the phlogiston, which was contained in all combustible bodies. It was liberated on burning organic matter, as well as in treating metals with heat (in air). That action transforms metals into what we now call oxides. An oxide, however, can reacquire its phlogiston by heating it with carbon, which, in itself is almost pure phlogiston. Stahl was aware that oxides weigh more than the metals from which they originate, but there were excuses for assigning to phlogiston a negative weight or other ways out of the difficulty.

The phlogiston theory could be bent to explain many facts, and it has its importance in the history of chemistry. Ultimately, phlogiston proved to be a sort of negative oxygen, if I may say so, oversimplifying the case.

Antoine-Laurent Lavoisier (1743–1794), by systematically using the balance in studying chemical reactions, concluded the downfall of phlogiston, but as a provisional hypothesis, phlogiston had its function. To give an idea of the popular fame of the theories of Stahl and Black, I will quote an ode, celebrating the first human flight and dedicated to R. Montgolfier by the Italian poet Vincenzo Monti in 1784.

Non mai Natura all'ordine	Never has Nature, subject to
delle sue leggi intesa	the order of its Laws
dalla potenza chimica	suffered such offense from
soffri più bella offesa.	the powers of chemistry.
mirabil'arte, ond'alzasi	Admirable art, that elevates
di Sthallio e Black la fama,	the fame of Stahl and Black,
pera lo stolto cinico	woe to the cynical fool
che frenesia ti chiama.	who calls you a phantasy!

The verses sound a little ridiculous, but they reflect the mood. The whole poem is stronger in mythology than in science.

The central question that confronted the natural philosophers was, What is heat? Could it be reduced to something else more primitive? Two doctrines held the field. According to one, heat was a substance, with or without weight; according to the other, heat was a mode of motion, possibly a vibration.

Robert Boyle (1627–1691) was a distinguished contemporary of Newton. He did much work in physics and chemistry, often collaborating with Hooke. His most important discovery is the relation between pressure and volume for a gas at constant temperature: $pV = constant$. In the background of the figure is the vacuum pump he used for his experiments.

The first view was based on the phenomena observed in mixing substances at different temperatures. The results could be accounted for by an indestructible extra substance called by Lavoisier heat or caloric. Latent (hidden) heat fitted into the picture by assuming that caloric could combine with material atoms of a substance and become latent, or be set free and become detectable by the thermometer.

The other view was supported by the obvious phenomena of heat development by friction. That view went back to ancient times, but had been eloquently maintained, for example, by Robert Boyle (1627–1691), who wrote as follows:

> If a somewhat large nail is driven by a hammer into a plank, it will receive diverse strokes on the head before it grows hot; but when it is driven to the head, so that it can go no farther, a few strokes will suffice to give it a considerable heat; for whilst at every blow of the hammer, the nail enters farther and farther into the wood, the motion that is produced is chiefly progressive, and is of the whole nail tending one way; whereas when the motion is stopped then the impulse given by the stroke, being unable either to drive the nail further on, or destroy its entireness, must be spent in making a various, vehement, and intestine commotion of the parts themselves, and in such an one we formerly observed the nature of heat to consist.

Another example, quoted by James Prescott Joule (1818–1889), is that of the philosopher John Locke (1632–1704), the friend of Newton, who similarly described heat's relation with friction: "Heat is very brisk agitation of the insensible parts of the object, which produces in us that sensation from whence we denominate the object hot; so what in our sensation is heat, in the object is nothing but motion."

The two views of heat, however, coexisted, and no contradiction between them was perceived for a long time. For example, Pierre-Simon Laplace (1749–1827) and Lavoisier in a memoir on heat, written in 1786, said:

> We will not decide at all between the two foregoing hypotheses. Several phenomena seem favorable to the one such as the heat produced by the friction of two solid bodies, for example; but there are others which are explained more simply by the other — perhaps both hold at the same time. . . . In general, one can change the first hypothesis into the second by changing the words *"free heat, combined heat* and *heat released"* into *vis viva,* loss of *vis viva* and increase of *vis viva.*

Important support for a mechanical origin of heat came from experiments conducted by Benjamin Thompson, Count Rumford (1753–1814), who was born in Woburn, Massachusetts. In the War of Independence, he sided with the British and left Boston in 1776, escaping to England, where he became an undersecretary in the Colonial Office. He had technical interests, and in London performed experiments on guns and explosives. His scientific prowess was recognized by election to the Royal Society in 1779. He later returned to America and commanded a regiment of the royal forces, but at the conclusion of peace, he returned to England and was put at half pay, although he was knighted in recognition of his services. From England he emigrated to Germany in the service of Elector Karl Theodor of Bavaria. There he worked on making cannons and on administrative tasks in gratitude for which he received the title of Count Rumford of the Holy Roman Empire. He founded the

Benjamin Thompson (1753–1814), later Count Rumford, was an American who had a most adventurous life. He emigrated to Europe, where he lived in England, Germany, and France, serving the respective governments, sometimes in dubious activities. His intuitions on the nature of heat and on the conservation of energy were significant. He was the founder of the Royal Institution in London.

English Garden, one of the prides of Munich, and he is honored by a statue in one of the main streets of the city. The cannon work had notable scientific importance. He showed that part of the work performed in boring a cannon went into heat, and using a blunt tool demonstrated that as long as work was supplied, heat was created. He even gave a rough estimate of the mechanical equivalent of 1 calorie for 5.5 joules. The correct value is 4.18 joules. The interpretation of his experiments remained controversial. Rumford, and later Thomas Young (1773–1829) maintained that they were incompatible with a material theory of heat and pointed to movement or vibrations as the cause of heat.

Scientific activity was only a small part of Rumford's life. He had been a spy for various nations, perhaps even a double agent. He was also an inventor, a social dreamer, and a philanthropist who used his fortune, obtained from a variety of sources (not all unimpeachable), in the founding of the Royal Institution at London, where he named Sir Humphry Davy (1775–1820) as director. Davy was only twenty-three years old, and the appointment shows keen judgment on the part of Rumford. He also instituted the Rumford Medal, one of the highest honors conferred by the Royal Society.

He then moved to Paris, where he was well received by Napoleon and Talleyrand and courted and married Lavoisier's widow. He did practical inventions in Paris, as he had in the past, and maintained contacts with the major scientists of the Académie Française, entering into arguments with Joseph Louis Lagrange (1736–1813) and Laplace. He died in Auteuil in 1814. Rumford is a typical figure of an eighteenth-century adventurer, and a detailed account of his life reads like a novel that might call to mind Casanova's memoirs.

Following Rumford's work, Davy performed an important experiment, casting doubt on the material nature of heat. He showed that two pieces of ice melt when rubbed against each other. The experiment was important because it was known that the specific heat of water is greater than that of ice, and that precluded an interpretation based on a material theory of heat.

Rumford and Davy's work did not, however, defeat the material theory of caloric. For instance, Jean Baptiste Joseph Fourier (1768–1830) wrote in 1822 his admirable book on heat conduction, in which he assumed that heat was an indestructible substance. He could account for all the phenomena of heat conduction he analyzed by that hypothesis. In a different context, Laplace corrected Newton's calculation of the velocity of sound by pointing out that the sound waves produce adiabatic (without transport of heat), not isothermal (constant temperature) compressions, as Newton had assumed. In that calculation of the adiabatic elasticity of a gas, Laplace, as well as Siméon-Denis Poisson (1781–1840), used the caloric theory and the result agreed with experiment. That success strengthened the caloric theory. Also, measurements on gas specific heats seemed to agree with this theory, although in reality they do not.

The doctrine of the convertibility of work into heat at a fixed rate was ultimately established around the middle of the nineteenth century, and with it came the conservation of energy, the first principle of thermodynamics and one of the major tenets of modern science. Before that time, however, the second principle had been discovered, and since the two principles are independent of each other, there was for a brief period a thermodynamics in which the second principle held good and the first did not. It was as if one had discovered noneuclidean geometry before the euclidean.

In a related field, the study of the laws of gases was of paramount importance for both chemistry and physics. It started at the time of Newton. One of the first results was the discovery of the relation between volume and pressure by Boyle in 1662 and l'Abbé Edme Mariotte (*d.* 1684) in 1679. Thus, the French call it Mariotte's law, while the rest of the world calls it Boyle's law. It simply states that at constant temperature, the volume of a gas is inversely proportional to the pressure. Boyle was the fourteenth child of the first Earl of Cork, and was born in 1627 at Lismore in Ireland. He traveled on the continent, as usual for aristocrats, and had his share of reverses and successes in a turbulent period of Irish and English history.

From 1655 to 1668, he settled in Oxford, where there was a strong group of natural philosophers, including the mathematician John Wallis (1616–1703), the architect of St. Paul's Cathedral in London, Sir Christopher Wren (1632–1723), Robert Hooke (1635–1702), and the anatomist Thomas Willis (1621–1675). Boyle hired Hooke as an assistant, impressed by his uncommon ability as an instrument maker, and together they built vacuum pumps in the wake of Otto von Guericke (1602–1686) of Magdeburg. In 1662 Boyle published a *Defense of the Doctrine Touching the Spring and Weight of the Air.* In it he describes how he made a U tube closed at one end:

> We began to pour Quicksilver into the long leg of the siphon, which by its weight pressing up that in the shorter leg, did by degrees streighten the included air, and continued this pouring in of Quicksilver till the air in the shorter leg was, by condensation, reduced to take up but half the space it possess'd (I say *possess'd* not *filled*) before . . . we observed that the Quicksilver in that longer part of the tube was 29 inches higher than the other.

The final conclusion was that he could verify the "hypothesis that supposes the pressures and expansions to be in reciprocal proportion." Note, however, that Boyle thought the elasticity of a gas was a static phenomenon, as though the substance contained many springs that at rest pressed on the walls. He did not attribute the pressure to molecular collisions against the walls. As an active member of the Royal Society, Boyle wrote much on many diverse subjects, from chemistry to medicine, but he is remembered chiefly for the gas law.

Thanks to the work of the chemists, many new gases became known in the eighteenth century, and it became possible to measure their physical properties. They all obeyed, approximately at least, Boyle's law. In the first years of the nineteenth century, other thermal properties of gases were studied, especially in France. The expansion coefficient at constant pressure for air was measured by Alessandro Volta (1745–1827) in 1791 and found to be $\frac{1}{273}$ on the Celsius scale. In 1802 Joseph Louis Gay-Lussac (1778–1850) found that this coefficient was the same for all gases. He also found the important and, for him, surprising fact that a gas expanding in a vacuum does not change temperature.

The specific heats at constant pressure and at constant volume were much more difficult to measure, but reasonably good values were obtained by François Delaroche and Jacques Étienne Bérard (1789–1869) around 1811. Those measurements, however, reported an erroneous dependence of the specific heat on pressure, an error that influenced Sadi Carnot (1796–1832).

The chemical evidence found an important interpretation in Amedeo Avogadro's (1776–1856) law (1811), stating that all gases at the same temperature and pressure contain the same number of molecules per unit volume. That fundamental law has deep implications for chemistry and for statistical mechanics that were fully grasped only about fifty years later. The same law was independently rediscovered by André-Marie Ampère (1775–1836) in 1814.

Henri Victor Regnault (1810–1878), initially an organic chemist from 1840 to 1860 (he then became a physicist), refined the measurements of many of those thermal magnitudes, furnishing sets of numbers unsurpassed for a long time. His Paris laboratory was a model for precise measurements of vapor pressures, specific heats, equations of state for real gases that slightly departed from Boyle's law, and so on, and he became the foremost authority in the field. That experimental material proved of great importance for the verification of the theoretical sciences of heat: thermodynamics and statistical mechanics. Furthermore, Regnault's experimental taste and style were propagated through many visitors and pupils from his laboratory to other European countries and had a significant influence in the formation of a whole generation of physicists.

Patriotism, Engineering, and Genius: Carnot and His Prophet William Thomson

In all didactical expositions of thermodynamics, conservation of energy, the first principle, properly precedes Sadi Carnot's principle, the second principle. However, as pointed out, the two are independent, and, historically, the idea of reversibility and the essentials of the second principle were grasped before the first principle had gained recognition.

The second principle of thermodynamics, and much more in the science of heat, is due to Sadi Carnot, a unique figure in science that reminds one, in the greatness of his accomplishments and in the brevity of his life, of his romantic mathematical contemporaries Niels Hendrik Abel (1802–1829) and Evariste Galois (1811–1832).

Carnot wrote only one scientific work — a slim book of one hundred and eighteen pages, entitled *Réflexions sur la puissance motrice de feu* (Reflections on the Motive Power of Fire). It appeared in 1824 in a printing of six hundred copies, done at the author's expense, and went almost unnoticed, although the author, the son of Lazare Carnot a very prominent Frenchman and member of the Académie des Sciences who had died the previous year, should have elicited some curiosity on the

Lazare Carnot (1753–1823), the "Organizer of Victory" and the father of Sadi. In his tempestuous career he was a scientist, military chief, and politician. He held a variety of positions that reflect the revolutionary times in which he lived. (Courtesy of the Library, University of California, Berkeley.)

part of his father's friends, who heard a brief description of the publication at one of the meetings of the Académie. The book, however, was studied by an alumnus of the École Polytechnique, Émile Clapeyron (1799–1864). Two years after graduation, Clapeyron went to Russia for eleven years to practice engineering and teach at the engineering school in St. Petersburg. On his return to France, he worked on railroad construction and was much interested in improving locomotives. In 1833 he came across Carnot's book, studied it, and reproduced its essential part in a more analytical form, publishing his results in the *Journal de l'École polytechnique* of 1834, after refusal by other journals. Carnot had been dead one year by then, and ten years had elapsed since the appearance of the *Réflexions*. It is from Clapeyron's paper that William Thomson, later Lord Kelvin (1824–1907), in 1849 became aware of the existence of Carnot's booklet, and recognized it for the fundamental work it is.

Sadi Carnot was born into a family notable in the history of France. He had a brother, Hippolyte, who became prominent in politics, and who, in turn, had a son, another Sadi, who was president of the French Republic from 1887 to 1894. His father, Lazare (1753–1823), was called the "Organizer of Victory" for his decisive part in the arming and organizing of the Revolutionary Army in 1794, when he was a member of the ruling Committee of Public Safety. As an ardent revolutionary, he had voted for the execution of the King. Later, he rallied to Napoleon, and in 1800 he was his minister of war. But Lazare Carnot was not only an administrator and a politician; he was also a mathematician and a scientist who wrote an important book on the calculus and has a theorem in trigonometry named after him. He was also interested in mechanics, studying both its theoretical foundations and practical problems, such as how to build efficient machines. He recognized that sudden changes in velocity in parts of a machine were detrimental and, in particular, that if one wanted to extract all possible useful mechanical work from falling water, the turbines had to be built so that water could enter them without a collision and leave them with no relative velocity. That type of thought may very well have influenced his son Sadi. Lazare had two children — Sadi, the physicist, and Hippolyte.

Sadi was educated mostly by his father until he was admitted to the École Polytechnique in a class of one hundred and seventy-nine young people (four hundred and seventeen had applied). He finished tenth in a class of sixty-five in 1814, when the Napoleonic Empire was tottering. Lazare served Napoleon once more during the Hundred Days before Waterloo. After Napoleon's final downfall, Lazare was exiled and died in Magdeburg.

Sadi held various positions in the army and as a professional engineer in public service. He retired from the army in 1828 as a captain. He had inherited a modest fortune from his maternal grandfather, and although he worked occasionally on engineering questions, he does not seem to have had a continuous occupation. In 1832 he became seriously ill with scarlet fever and was interned in a hospital. He recovered from that disease, but shortly thereafter contracted cholera in an epidemic then ravaging Paris; he died on August 24, 1832.

Around 1824, at the age of twenty-eight, Carnot had become interested in steam engines. They had been invented and developed by practical men for practical purposes. The development had been mostly empirical. We admire the ingenuity and resourcefulness of James Watt (1736–1819), but his attitude was more that of an engineer than that of a scientist. Great Britain was then ahead of the world in technical proficiency. The Industrial Revolution had started there and had allowed the relatively small island to become the world's first power, and Carnot was very sensitive to these events. His French patriotism would have liked to see his country reach and possibly surpass England, which had not long before triumphed over Napoleon. His contribution to those patriotic aims was to be a scientific study of the steam engine.

Toward the beginning of E. Mendoza's 1960 translation of *Réflexions*, we find the following:

The most signal service that the steam-engine has rendered to England is undoubtedly the revival of the working of the coal mines, which had declined, and threatened to cease entirely, in consequence of the continually increasing difficulty of drainage, and of raising the coal. We should rank second the benefit to iron manufacture, both by the abundant supply of coal substituted for wood just when the latter had begun to grow scarce, and by the powerful machines of all kinds, the use of which the introduction of the steam-engine has permitted or facilitated.

Iron and heat are, as we know, the supporters, the bases, of the mechanic arts. It is doubtful if there be in England a single industrial establishment of which the existence does not depend on the use of these agents, and which does not freely employ them. To take away today from England her steam-engines would be to take away at the same time her coal and iron. It would be to dry up all her sources of wealth, to ruin all on which her prosperity depends, in short, to annihilate that colossal power. The destruction of her navy, which she considers her strongest defence, would perhaps be less fatal. . . .

If the honor of a discovery [of the steam engine] belongs to the nation in which it has acquired its growth and all its developments, this honor cannot be here refused to England. Savery, Newcomen, Smeaton, the famous Watt, Woolf, Trevithick, and some other English engineers, are the veritable creators of the steam-engine. It has acquired at their hands all its successive degrees of improvement. Finally, it is natural that an invention should have its birth and especially be developed, be perfected, in that place where its want is most strongly felt.

Notwithstanding the work of all kinds done by steam-engines, notwithstanding the satisfactory condition to which they have been brought today, their theory is very little understood, and the attempts to improve them are still directed almost by chance.

Carnot wanted to understand the fundamentals, hoping they would help improve the practice. He thus started to try to schematize the steam engine in the most abstract way. He was guided by an analogy with a water turbine, where water falls from an upper to a lower level and its potential energy is transformed into

Sadi Carnot (1796–1832) at age seventeen as a pupil of the École Polytechnique and at age thirty-four, about the time he wrote Réflexions sur la puissance motrice du feu (1824). *(Courtesy of the Library, University of California, Berkeley.)*

"motive power." For a steam engine, he thought, heat falls from a higher temperature to a lower one and, in so doing, delivers motive power. Carnot, at the time, followed the caloric theory, and so missed the fundamental difference between heat and water: Whereas the amount of water is constant, the heat delivered at the lower temperature is diminished by an amount proportional to the work obtained. He did not know that fact, and he proceeded under the assumption that the heat was conserved (caloric). Many of the conclusions he reached, however, are easily extended to the real case, and it is his way of thinking that is of paramount importance.

He addressed the fundamental question: Is there a maximum work obtainable from heat falling from one temperature to the other, and, if so, on what does it depend?

In a steam engine, for example, heat is conferred to steam contained in a cylinder closed by a piston. It stands to reason that the heat should be conferred with as little temperature jump as possible, because in a turbine water has to enter without collision, and, furthermore, when heat goes to a lower temperature by conduction, no work is obtained. In developing that concept, Carnot saw that all transformations in the machine had to be a succession of equilibrium states. As a consequence, however, the machine could function in one direction, or, by reversing all the operations, in the opposite. The machine had to be *reversible*.

The machine must have a source of heat at high temperature, a sink of heat at lower temperature, and a system that transfers the heat, for example, a cylinder with a piston. It is important that the transformation in which the heat passes from the higher to the lower-temperature reservoir leaves the transmitting system completely unchanged. Such an operation was described by Carnot and is called cyclic, meaning by this that after a certain set of operations, the transmitting system returns to its original state. Naturally, this is not the case for the reservoirs unless they are maintained somehow at constant temperatures, as Carnot assumed. Here are Carnot's words describing his famous cycle as applied to an air machine:

. . . Let us imagine an elastic fluid, atmospheric air for example, shut up in a cylindrical vessel, *abcd,* provided with a movable diaphragm or piston, *cd.* Let there be also two bodies, A and B, kept each at a constant temperature, that of A being higher than that of B. Let us picture to ourselves now the series of operations which are to be described:

(1) Contact of the body A with the air enclosed in the space *abcd* or with the wall of this space — a wall that we will suppose to transmit the caloric readily. The air becomes by such contact of the same temperature as the body A; *cd* is the actual position of the piston.

(2) The piston gradually rises and takes the position *ef.* The body A is all the time in contact with the air, which is thus kept at a constant temperature during the rarefaction. The body A furnishes the caloric necessary to keep the temperature constant.

(3) The body A is removed, and the air is then no longer in contact with any body capable of furnishing it with caloric. The piston meanwhile continues to move, and passes from the position *ef* to the position *gh.* The air is rarefied without receiving caloric, and its temperature falls. Let us imagine that it falls thus till it becomes equal to that of the body B; at this instant the piston stops, remaining at the position *gh.*

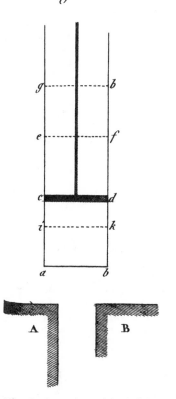

The Carnot cycle as described by him. The figure shows a frictionless cylinder with a thermally isolated piston. The bottom of the cylinder is thermally conducting and either in contact with a hot reservoir (A) or a cold reservoir (B). (From Réflexions sur la puissance motrice du feu, *1824).*

(4) The air is placed in contact with the body *B*; it is compressed by the return of the piston as it is moved from the position *gh* to the position *cd*. This air remains, however, at a constant temperature because of its contact with the body *B*, to which it yields its caloric.

(5) The body *B* is removed, and the compression of the air is continued, which being then isolated, its temperature rises. The compression is continued till the air acquires the temperature of the body *A*. The piston passes during this time from the position *cd* to the position *ik*.

(6) The air is again placed in contact with the body *A*. The piston returns from the position *ik* to the position *ef*; the temperature remains unchanged.

(7) The step described under number (3) is renewed, then successively the steps (4), (5), (6), (3), (4), (5), (6), (3), (4), (5); and so on.

In these various operations the piston is subject to an effort of greater or less magnitude, exerted by the air enclosed in the cylinder; the elastic force of this air varies as much by reason of the changes in volume as of changes of temperature. But it should be remarked that with equal volumes, that is, for the similar positions of the piston, the temperature is higher during the movements of dilatation than during the movements of compression. During the former the elastic force of the air is found to be greater, and consequently the quantity of motive power produced by the movements of dilatation is more considerable than that consumed to produce the movements of compression. Thus we should obtain an excess of motive power — an excess which we could employ for any purpose whatever. The air, then, has served as a heat-engine; we have, in fact, employed it in the most advantageous manner possible, for no useless re-establishment of equilibrium has been effected in the caloric.

All the above-described operations may be executed in an inverse sense and order. Let us imagine that, after the sixth period, that is to say the piston having arrived at the position *ef*, we cause it to return to the position *ik*, and that at the same time we keep the air in contact with the body *A*. The caloric furnished by this body during the sixth period would return to its source, that is, to the body *A*, and the conditions would then become precisely the same as they were at the end of the fifth period. If now we take away the body *A*, and if we cause the piston to move from *ik* to *cd*, the temperature of the air will diminish as many degrees as it increased during the fifth period, and will become that of the body *B*. We may evidently continue a series of operations the inverse of those already described. It is only necessary under the same circumstances to execute for each period a movement of dilatation instead of a movement of compression, and reciprocally.

The result of these first operations has been the production of a certain quantity of motive power and the removal of caloric from the body *A* to the body *B*. The result of the inverse operations is the consumption of the motive power produced and the return of the caloric from the body *B* to the body *A*; so that these two series of operations annul each other, after a fashion, one neutralizing the other.

In simpler, graphical language, I present a figure used first by Clapeyron, in which Carnot's cycle is illustrated in a *p–V* diagram. The cycle is composed of four transformations—two adiabatic and two isothermal.

The machine is clearly reversible. We can now couple two reversible machines, one working in the direct sense and one in the opposite. If one had a higher efficiency than the other, we could realize a perpetuum mobile, which Carnot holds for an absurdity. Here are his words:

> The operations which we have just described might have been performed in an inverse direction and order. . . .
>
> By our first operations there would have been at the same time production of motive power and transfer of caloric from the body *A* to the body *B*. By the inverse operations there is at the same time expenditure of motive power and return of caloric from the body *B* to the body *A*. But if we have acted in each case on the same quantity of vapor, if there is produced no loss either of motive power or caloric, the quantity of motive power produced in the first place will be equal to that which would have been expended in the second, and the quantity of caloric passed in the first case from the body *A* to the body *B* would be equal to the quantity which passes back again in the second from the body *B* to the body *A*; so that an indefinite number of alternative operations of this sort could be carried on without in the end having either produced motive power or transferred caloric from one body to the other.

Émile Clapeyron (1799–1864), French engineer and student of the École Polytechnique. He understood the importance of Carnot's work and, through his reformulation, rescued it from oblivion. He was also prominent as a railroad and bridge builder and for his teaching in Russia. (Courtesy of the Archives de l'Académie des Sciences de Paris.)

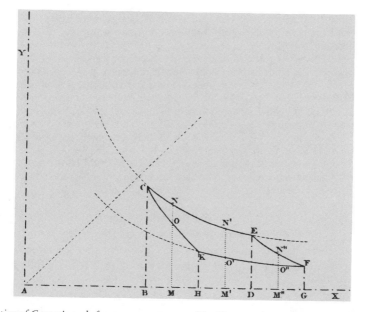

Illustration of Carnot's cycle for a gas, as represented by Clapeyron in a paper (Journal de l'École Polytechnique XXIII:190, 1834). *Abscissa is volume; ordinate is pressure. Lines CE and KF are isothermals; lines CK and EF are adiabatics. The area of the curvilinear rectangle CEFK is the mechanical work done in a cycle. (Courtesy of the Library, University of California, Berkeley.)*

Now if there existed any means of using heat preferable to those which we have employed, that is, if it were possible by any method whatever to make the caloric produce a quantity of motive power greater than we have made it produce by our first series of operations, it would suffice to divert a portion of this power in order by the method just indicated to make the caloric of the body B return to the body A from the refrigerator to the furnace, to restore the initial conditions, and thus to be ready to commence again an operation precisely similar to the former, and so on: this would be not only perpetual motion, but an unlimited creation of motive power without consumption either of caloric or of any other agent whatever. Such a creation is entirely contrary to ideas now accepted, to the laws of mechanics and of sound physics. It is inadmissible.* We should then conclude that *the maximum of motive power resulting from the employment of steam is also the maximum of motive power realizable by any means whatever.*

The important footnote adds:

* The objection may perhaps be raised here, that perpetual motion, demonstrated to be impossible by mechanical action alone, may possibly not be so if the power either of heat or electricity be exerted; but is it possible to conceive the phenomena of heat and electricity as due to anything else than some kind of motion of the body, and as such should they not be subjected to the general laws of mechanics? Do we not know besides, *a posteriori,* that all the attempts made to produce perpetual motion by any means whatever have been fruitless? — that we have never succeeded in producing a motion veritably perpetual, that is, a motion which will continue forever without alteration in the bodies set to work to accomplish it? The electromotor apparatus (the pile of Volta) has sometimes been regarded as capable of producing perpetual motion; attempts have been made to realize this idea by constructing dry piles said to be unchangeable; but however it has been done, the apparatus has always exhibited sensible deteriorations when its action has been sustained for a time with any energy.

The general and philosophic acceptation of the words *perpetual motion* should include not only a motion susceptible of indefinitely continuing itself after a first impulse received, but the action of an apparatus, of any construction whatever, capable of creating motive power in unlimited quantity, capable of starting from rest all the bodies of nature if they should be found in that condition, of overcoming their inertia; capable, finally, of finding in itself the forces necessary to move the whole universe, to prolong, to accelerate incessantly, its motion. Such would be a veritable creation of motive power. If this were a possibility, it would be useless to seek in currents of air and water or in combustibles this motive power. We should have at our disposal an inexhaustible source upon which we could draw at will.

Thus, all reversible machines must have the same efficiency, that being defined as the ratio of the work obtained W to the heat H given by the hot source. Since no mention is made of the substance used in the cycle, it is clear that the substance is not important. The essential point is the reversibility of the cycle. Also, it follows that the efficiency of a reversible machine can depend only on the temperature of the hot and the cold source. All this is true and is the essence of the great discoveries of Carnot.

In *Réflexions,* Carnot made the assumption that the heat delivered at the cold source H' is the same as H. That, of course, is an error, because in reality $H - H' = W$ by the conservation of energy. Carnot, in his published work, made the wrong assumption of the conservation of heat. That he was suspicious of it is demonstrated by the footnote quoted above. It is most remarkable, however, that the grievous flaw left unchanged the major part of his deductions, and that when the doctrine of the conservation of energy was established (about twenty years after the publication of his book and fifteen years after his death), the *Réflexions* could be suitably changed and remained a pillar of the new science of thermodynamics.

Carnot had had clear premonitions of the conservation of energy. His early death possibly prevented his bringing his thoughts to fruition. Unfortunately, the notes he left were not comprehended by his brother, who published them only in 1878, when thermodynamics was already firmly established. Thus, Carnot's thoughts on the conservation of energy did not influence the work of his successors, but it is, nevertheless, worthwhile to quote from his manuscripts, which clearly show how far he had come. Some notes found among them are:

> According to some ideas which I have formed on the theory of heat, the production of a unit of motive power necessitates the destruction of 2.70 units of heat. [The correct value in his units is 2.39.]
> A machine which would produce 20 units of motive power per kilogram of coal ought to destroy $20 \times 2.70/7000$ of the heat developed by the combustion. $20 \times 270/7000 = 8/1000$ roughly, that is less than $1/100$.

> When a hypothesis no longer suffices to explain phenomena, it should be abandoned.
> This is the case with the hypothesis which regards caloric as matter, as a subtle fluid.

The experimental facts tending to destroy this theory are as follows:

(1) The development of heat by percussion or the friction of bodies (Rumford's experiments, friction of wheels on their spindles, on the axles, experiments to be made). Here the elevation of temperature takes place at the same time in the body rubbing and in the body rubbed. Moreover, they do not change perceptibly in form or nature (to be proved). Thus heat is produced by motion. If it is matter, it must be admitted that the matter is created by motion.

Fourteen years after Clapeyron, Carnot's work was continued and expanded by William Thomson, whom we have met previously many times. He is one of the great figures of the century. He could have been placed with equal justification among the electricians or the mathematical physicists. In his long life, he came to be considered as the first physicist of the British Empire, was equally admired by the rest of the world, and was honored in all possible fashions. He deserved all this, but it was made easier by his long life and his practical endeavors that brought him not only fame, but also wealth and popularity outside his profession.

William Thomson was the son of James Thomson, who, from rather humble origins, had succeeded to become a well-known professor of mathematics, whose textbooks went through many editions. Of his seven children, one (also James) became a noted scientist and engineer, while William, the second male, became the famous physicist. Both parents were of Scottish origin, but for some years lived in Belfast, Ireland, where William was born. In 1832 James Thomson was appointed professor of mathematics at Glasgow University. His young sons were permitted to attend his lectures when they were little more than children, and they matriculated at the university at ages twelve and ten, respectively. At age fifteen, William was reading Lagrange's *Mécanique analytique* and Fourier's *Théorie analytique de la chaleur;* the latter made a deep impression on Thomson and became his constant companion and favorite reading.

William Thomson at age twenty-eight. (Courtesy of AIP Niels Bohr Library, Zeleny Collection.)

The Thomsons prospered in a modest way, because by 1839 the father took the whole family on a tour of the continent of Europe. Needless to say, a prodigy such as William had already learned German and French, in addition to the classical languages. By 1841, when he was seventeen, William was ready to go to Cambridge, at Peterhouse, the traditional Scottish college. William rapidly became well known in Cambridge. He published some mathematical papers, developed musical tastes, and, on the whole, enjoyed life. Over three years his father sent him £774, and he wrote to his son, "How is this to be accounted for? Have you lost money or been defrauded of it, or have you lived on a more expensive scale? Do consider the matter and state fully and clearly what you take to be the reason." The stern words may conceal that the father was immensely solicitous for the welfare and the career of his dear son. In 1841 William took the tripos and placed second. It was a disappointment because everybody expected him to be the senior wrangler. Nevertheless, shortly thereafter he was bound for Paris to complete his education. In the meantime he had discovered a most important paper, *Essay on the Application of Mathematical Analysis to the Theories of Electricity and Magnetism* by George Green (1793–1841), which contained, among other things, his famous reciprocity theorems. The *Essay* was almost unknown and extremely hard to find. Thomson rapidly assimilated its content and put it to good use.

In Paris the young physicist became acquainted with many of the mathematical lights of the French capital, including Friedrich Otto Rudolf Sturm (1841–1919), Augustin-Louis Cauchy (1789–1857), and Joseph Liouville (1809–1882). Liouville was all excited when he heard that Thomson had Green's essay, because he had independently found some of Green's results, but had never seen Green's paper. At the Sorbonne, Thomson attended lectures by Jean-Baptiste-André Dumas (1800–1884), the chemist, and by Regnault, the physicist, at the Collège de France. He volunteered to help the latter, and started to work regularly with him as a laboratory assistant. Then he became acquainted with Clapeyron's paper, mastered it, and was naturally very anxious to obtain a copy of Carnot's book. Thomson tells how he went to many booksellers and *bouquinistes* only to find the works of Lazare or Hippolyte Carnot. Nobody had ever heard of Sadi. It was only in 1848 that he saw for the first time a copy of *Réflexions*.

In 1846 the Glasgow Professor of Natural Philosophy died. It was not an unexpected event, and the Thomsons, father and son, had been carefully preparing the succession in favor of William. Their perfectly legitimate efforts are reflected in many letters. In 1846, when he was appointed, William Thomson was twenty-two years old. Among those who recommended him were George Gabriel Stokes (1819–1903), Augustus De Morgan (1806–1871), and Regnault. He held the job for fifty-three years. The university had a tradition of applied research, had counted James Watt among its instrument makers and Joseph Black of specific heat fame had served there. The collection of instruments contained models of old steam machines of the Thomas Newcomen (1663–1729) type, as well as modern instruments. The new professor started by organizing his teaching. For example, he adapted a basement for laboratory use where his students could do practical work, often connected with Thomson's personal research. That laboratory was one of the first of its kind. In time, it expanded and became a valid help to Thomson's electrical engineering semicommercial enterprises.

Thomson himself lectured twice a day four days each week, including recitations and laboratory. His lectures were brilliant but difficult, and many students seem to have enjoyed more the personality of the professor than to have followed his mathematics and digressions. Thomson demonstrated many spectacular experiments that impressed the students, but during the lectures his rapid and fertile mind would take hold of a new subject, often forgetting the main argument. On the whole, however, his teaching in the classroom, through example and through his books, was very effective, and many leaders in electrical engineering came from Glasgow. Related to his teaching was the famous text written in collaboration with the Edinburgh Professor Peter Guthrie Tait's (1831–1901) *Treatise on Natural Philosophy*. It was never finished, and the two volumes we have are limited to mechanics. Today the book appears badly dated. Fifty years ago, Enrico Fermi suggested it to me when I asked him for a good text to study. I suspect he knew it only superficially; I found it difficult and unrewarding. Nevertheless, Thomson's contemporaries compared it to Newton's *Principia*. Maxwell reviewed it analytically, and in reading his review, I discerned some of his sharp but well-placed irony. The Glasgow chair was extremely congenial to Thomson. Offers from many other universities and institutions for exalted positions never tempted him away.

Thomson in 1847 had studied Clapeyron's paper, and he saw that it provided a way of building a scale of temperatures independent of any thermometric substance, an absolute temperature scale. Thomson's idea was to define the temperature through a Carnot engine, which is strictly independent of the fluid. A certain quantity of heat falling by one degree, as newly defined, gives a certain amount of work. An imperfect example of such a scale is given by a gas thermometer. If the gas laws were strictly obeyed under all circumstances, all gas thermometers would give the same temperature (this is not an accident), but in practice even gas thermometers are not absolute.

Thomson, in a paper of 1848, adhered with Carnot to the hypothesis of the conservation of heat and thus assumed that the amount of heat entering the engine at the higher temperature was the same as the amount leaving it at the lower temperature. In his words:

> The characteristic property of the scale I now propose is, that all degrees have the same value; that is, that a unit of heat descending from body A at the temperature $T°$ of this scale, to a body B at the temperature $T° - 1$, would give out the same mechanical effect whatever the number $T°$. This may justly be termed an absolute scale, since its characteristic is quite independent of the physical properties of any specific substance.

Of course, the foundation of the scale was wrong, because the quantity of heat supplied by A is not equal to that received by B, but the method could easily be adapted a few years later to the conservation of energy. It is remarkable that when Thomson was publishing his paper, he was already acquainted with the work of

James Prescott Joule (1818–1889).

Joule, shaking the foundations of his arguments. In a footnote to his paper on the absolute temperature, he says:

> This opinion [the conservation of heat], seems to be nearly universally held by those who have written on the subject. A contrary opinion, however, has been advocated by Mr. Joule of Manchester; some very remarkable discoveries which he has made with reference to the generation of heat by the friction of fluids in motion, and some known experiments with magnetoelectric-machines seeming to indicate an actual conversion of mechanical effects into caloric. No experiment, however, is adduced in which the converse operation is exhibited; but it must be confessed that as yet much is involved in mysteries with reference to these fundamental questions of natural philosophy.

In fact Thomson had met Joule, six years his senior, at a meeting of the British Association at Oxford in 1847 and learned from him of his experiments on the transformation of work into heat at a fixed rate. A certain amount of work was *equivalent* to a corresponding amount of heat, and the coefficient of reduction was a constant.

Many years later Thomson recalled:

> I can never forget the British Association at Oxford in 1847, when in one of the sections I heard a paper read by a very unassuming young man, who betrayed no consciousness in his manner that he had a great idea to unfold. I was tremendously

struck with the paper. I at first thought it could not be true, because it was different from Carnot's theory, and immediately after the reading of the paper I had a few words with the author, James Joule, which was the beginning of our forty years' acquaintance and friendship. On the evening of the same day, that very valuable institution of the British Association, its conversazione, gave us opportunity for a good hour's talk and discussion over all that either of us knew of thermodynamics. I gained ideas which had never entered my mind before, and I thought I, too, suggested something worthy of Joule's consideration when I told him of Carnot's theory. Then and there in the Radcliffe Library, Oxford, we parted, both of us, I am sure, feeling that we had much more to say to one another and much matter for reflection in what we had talked over that evening.

A collaboration in theoretical and experimental work soon followed. It became apparent that the old experiment of Gay-Lussac, showing that the expansion of a gas in a vacuum did not entail a change in temperature, was of paramount importance, and together the two friends improved on it. The ideas leading to the absolute scale of temperatures thus had to be modified, and Thomson proceeded to do it. The final result was that, as Carnot had proved, the efficiency of his engine, defined as the ratio of the work obtained W to the heat supplied at the hot reservoir Q, was still a function of the temperatures of the hot and cold reservoirs (T, T') only. The mechanical work, however, was the difference between Q and Q', the heat surrendered at the cold source. It was convenient to define the ratio T/T' by making it equal to Q/Q'. It followed that the efficiency $\dfrac{Q - Q'}{Q} = \dfrac{W}{Q} = \eta$ is given by $\eta = (T - T')/T$, and that there has to be an absolute zero because the efficiency cannot exceed 1.

The absolute scale of temperatures, thus defined, had to be compared with practical scales, and that was done in the next few years. For a perfect gas obeying the equation of state $pV = R\theta$, which defines the temperature θ, and the caloric condition expressed by the result of the Gay-Lussac experiment, the two scales coincide and are completely defined, once the size of the degree is established. The usual convention is that there are 100 degrees (centigrades) between the temperature of melting ice and boiling water, at standard atmospheric pressure. For real gases and other substances, Joule and Thomson established points of reference for obtaining absolute temperatures. At present, however, it is preferred to define the absolute temperature scale by assigning to the triple point of water the absolute temperature of 273.16 degrees (Kelvin), *exactly*. At the triple point, ice, vapor, and liquid are in equilibrium, and pressure as well as temperature are fixed. The scale coincides within a few hundredths of a degree with the scale described above.

Thomson also made other applications of thermodynamics to natural phenomena, including electrical ones, and became quite proficient in the art. One of the earliest had been done by his brother James and also by Clapeyron — finding the effect of pressure on the melting point of ice.

This great work may well be Thomson's most permanent contribution to physics, although it had been preceded by similar studies of Rudolf Clausius (1822 – 1888), which will be discussed later. The two investigations were indepen-

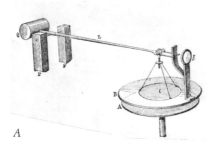

A

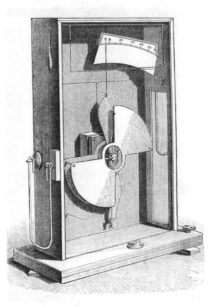

B

Three instruments from Kelvin's laboratory.
Figure A *is an absolute electrometer with*
guard ring. The difference of electrical poten-
tial between two disks is measured in absolute
electrostatic units by their mechanical attrac-
tion. The force $F = SV^2/8\pi d^2$ *is measured by*
a weight and a balance.
S is the surface of the
disk c, V the potential
difference, and d the
distance between the
disks. Figure B *is an*
electrometer for practical
laboratory use. I used one
like this for nuclear work
as late as 1934. Figure
C *is a galvanometer that*
pushes current sensitiv-
ity to the limit for its
times (1880). About
10^{-11} *amps are easily*
detectable. The instru-
ment contains three
magnetized needles at
the centers of two coils
and is arranged to mini-
mize the action of the
terrestrial magnetic
field, which is further
reduced by the large
compensating magnet
bar at the top.

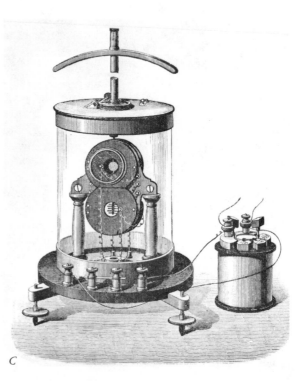

C

dent and both were of capital importance. The subject is perhaps somewhat recondite; Thomson's contemporaries have remembered and honored him for other accomplishments.

In many papers on electrical subjects, from electrostatics to transient currents, he treated diverse problems with superb craftsmanship; he revealed analogies between Fourier's theory of heat conduction and potential theory; he discussed aspects of Faraday's ideas on the propagation of electrical action; and he analyzed oscillating circuits and the alternate currents they generate. His papers influenced James Clerk Maxwell, who asked him for advice and for permission to work on the same lines of inquiry. In 1854 Maxwell told his father that Thomson answered that he was "very glad that I should poach on his electrical preserves."

Thomson also applied his physics to an entirely different field. He investigated possible origins of solar heat and the temperature balance of the earth. His methods were sound and interesting, but since he did not know the nuclear origin of the energy in the sun and in the earth, he could not possibly arrive at correct results. He tried to explain the origin of solar heat by meteorites falling in the sun or by gravitational contraction, but he estimated, in about 1854, the sun's "age" to be less than 5×10^8 years — at least ten times younger than that now accepted.

From the temperature gradient near the surface of the earth, he tried to reconstruct its thermal history and age. This estimate was also far too low: 4×10^8 years, while the answer is about 5×10^9 years. The geologists, basing their theories on rates of geological phenomena, immediately found fault with his estimates. They could not refute Thomson's mathematics, but they concluded his assumptions must be wrong. Also, the biologists could not fit the recent evolutionary ideas into the time-scale suggested by Thomson. The controversy lasted for many years, and Thomson was rather impervious to their correct objections. Ultimately, the discovery of radioactivity and of nuclear reactions showed the fault in Thomson's premises.

In 1855 he met Hermann von Helmholtz (1821 – 1894) for the first time, and they became lifelong personal friends. His German colleague describes him to his wife as follows:

> I expected to find him, who is one of the best mathematical physicists of Europe, a man slightly older than myself. I was quite surprised when a very young-looking fellow with light blond hair and a very feminine aspect came towards me. He had rented a room for me in the vicinity of his place and I had to transfer my things from the hotel to stay there. He is at Kreuznach because of his wife, who appeared shortly after him. She is a very charming and intelligent woman but very sick. He anyway surpasses all the great scientists I know personally as far as sharpness, clarity and mobility of mind is concerned. So much so that I feel next to him somewhat of a blockhead.

Thomson was then thirty-one years old and had already written the best ninety papers of the six hundred and sixty-one that he was to publish in his long career. He was also about to enter a new phase of his life. Electrical telegraphs had been operating for many years over increasing distances, and it was becoming realistic as a profitable business enterprise to plan a transatlantic cable that would link Europe to America. After preliminary studies, several tycoons organized the Atlantic Telegraph Company capitalized at £350,000. Thomson was nominated by the Scottish shareholders to serve as a director on a board of eighteen.

He was the only director with a deep technical understanding of the problems facing the enterprise. In previous years, he had developed the theory of signal transmission, governed by the "telegraphist equation." He knew how a signal propagates on a cable, and he understood the importance of the conductivity of the copper, of the dielectric constant of the insulator, and of the losses along the cable. His other technical colleague, E. O. W. Whitehouse, was an empirical self-taught man with a limited understanding of the problems.

The practical telegraph work changed Thomson's life. He was becoming a businessman and was assuming responsibilities and functions very different from those of academic life. It turned out that his personality was excellently suited to the new career, in that he combined his mastery of theory with superior engineering and inventive skills. He produced an uninterrupted stream of instruments adapted to current problems, among them the mirror galvanometer, in which a weightless beam of light replaces a pointer. Maxwell wrote a poem about it: I give it as an example of Maxwell's rhymes.

VALENTINE BY A TELEGRAPH CLERK ♂ TO A TELEGRAPH CLERK ♀.

"The tendrils of my soul are twined
　　With thine, though many a mile apart,
And thine in close-coiled circuits wind
　　Around the needle of my heart.

"Constant as Daniell, strong as Grove,
　　Ebullient through its depths like Smee,
My heart pours forth its tide of love,
　　And all its circuits close in thee.

"O tell me, when along the line
　　From my full heart the message flows,
What currents are induced in thine?
　　One click from thee will end my woes."

Through many an Ohm the Weber flew,
　　And clicked this answer back to me, —
"I am thy Farad, staunch and true,
　　Charged to a Volt with love for thee."

In another poem Maxwell again praises Thomson's invention:

LECTURES TO WOMEN ON PHYSICAL SCIENCE.

　　PLACE. — *A small alcove with dark curtains.*
　　　The Class consists of one member.

SUBJECT. — *Thomson's Mirror Galvanometer.*

The lamp-light falls on blackened walls,
   And streams through narrow perforations,
The long beam trails o'er pasteboard scales,
   With slow-decaying oscillations.
Flow, current, flow, set the quick light-spot flying,
Flow current, answer light-spot, flashing, quivering, dying.

O Look ! how queer ! how thin and clear,
   And thinner, clearer, sharper growing
The gliding fire ! with central wire,
   The fine degrees distinctly showing.
Swing, magnet, swing, advancing and receding,
Swing magnet! Answer dearest, What's your final reading?

O love ! you fail to read the scale
   Correct to tenths of a division.
To mirror heaven those eyes were given,
   And not for methods of precision.
Break contact, break, set the free light-spot flying;
Break contact, rest thee, magnet, swinging, creeping, dying.

Thomson's practical work in telegraphy started in 1856 with the laying of an unsatisfactory cable, which broke after 330 nautical miles. Thomson was on board the vessel that laid the cable. He studied the mechanical problems connected with the cable-laying operation and the electrical problems connected with the signal transmission. It would have been difficult to find a better person to investigate such questions of scientific engineering in which only practical approaches were insufficient, but strict contact with reality had to be maintained.

The year 1858 saw the next attempt at laying a new cable. Thomson was again aboard the ship. The expedition met a severe storm that imperiled the operation but ultimately, on August 5, 1858, Thomson sent the first signals. The cable lasted only about a month. The fault was in part with the construction of the cable and in part with the high voltages that Whitehead had forced on it. Several years had to pass before a new attempt. By 1865 the company had a cable of much better quality and also a much larger ship to carry it. Thomson was once more on board. The cable, however, snapped in mid-Atlantic, and the company had to suspend operations. Finally in 1866 the fourth attempt was successful. It was also possible to retrieve the broken cable of the previous year, so that in the end there were two cables. The event was celebrated; Thomson was knighted and received the freedom of the City of Glasgow.

Through his work on the cable and through his inventions, Sir William also became wealthy. His instruments were built in the workshops of James White, a Glasgow optician and instrument maker with whom he established a company. The firm prospered and, in due course, became a leader in the field. The company was reorganized several times, under various names, and exists today under the name of the Kelvin Hughes Division of Smith Industries.

Sir William Thomson's yacht, Lalla Rookh. *(Courtesy AIP Niels Bohr Library.)*

To complete the picture, I should mention the similar efforts of Ernst Werner von Siemens (1816–1892) and his German associates. In a nutshell, Great Britain was in the vanguard for science, but Cambridge dons did not relish practical applications; nor did the engineers put the necessary effort into acquiring scientific foundations. Thus, science and technology were much farther apart in Britain than in Germany. Siemens, the great German electrical engineer, commented explicitly on this around 1860, and as a businessman maintained that technology with a strong scientific basis would have competitive advantages. Siemens had a brother, Sir Charles William Siemens (1823–1883), who became a British subject and a great British engineer and industrialist; his business relations were international, and he was an astute observer of the situation. The two brothers were very conscious of the relations between science and technology and profited handsomely from their openmindedness. For further information the interested reader can look at the fascinating autobiography of Siemens mentioned in the bibliography.

In the middle of his very active life, on Christmas Eve 1860, Thomson fell and broke his left leg while playing on ice. The fracture was very serious and left him with a permanent limp, which somewhat impaired his mobility. Later, in 1870, he lost his wife after eighteen years of marriage. She had been in poor health most of her life, and her death was not unexpected; nevertheless, the event saddened him deeply. A picture taken at that time shows a face in deep sorrow.

In 1871 Sir William acquired a 126-ton sailing yacht and called it *Lalla Rookh,* after an 1817 poem by Thomas Moore. From then on he spent a sizeable

Netherall, the Thomson house in Largs, was built in 1882 near the sea to give him easy access to his yacht. The house was one of the first to be equipped with electric light.

amount of time on board. He often invited friends; we see him planning a cruise with Helmholtz, Maxwell, Thomas Henry Huxley (1825 – 1895), John Tyndall (1820 – 1893), and Tait, but unfortunately not all of them could go. On board he worked actively on physico-mathematical problems, frequently inspired by accidents of navigation. He always carried with him his famous green notebooks, and he filled them with calculations. Apparently, he competed with his friends, for example, Helmholtz, on who could solve problems faster.

Hydrodynamics became one of Thomson's favorite subject, especially vortex theory, where he found valuable theorems inspired by the work of Helmholtz. On the other hand, his speculations on vortex atoms fell far from the mark. One fruit of his navigations was the invention in 1876 of a special compass suitable for iron ships. Ultimately, it was adopted by the British Navy and held the field until it was replaced by modern gyroscopic compasses. Thomson's enterprises built many magnetic compasses and sounding apparatus at handsome profits.

Prompted by his practical experience and fortified by his theoretical knowledge, Thomson felt a strong need for a systematization of electrical measurements. Of course, he was not the only one. The French Revolution had made a giant step by introducing the metric system, but electrical measurements raised entirely new problems. The theoretical foundation for a system of absolute units had been laid by Carl Friedrich Gauss (1777 – 1855) and Wilhelm Edward Weber (1804 – 1891). "Absolute" means that they do not depend on specific substances or standards, but that they are based on the universal laws of physics. For electricity there are several systems of absolute units (see page 165). The practical realization of standards calibrated on the absolute units, the choice of suitable multiples of the absolute units for industrial use, and persuading the scientific-technical community to accept them were all important and difficult tasks. The British Association in 1861 had started this work by appointing a committee, of which Thomson was a member. They worked actively for years, and by 1881 an International Congress at Paris, dominated by Thomson and Helmholtz, and another Congress in 1893 in Chicago, accepted the new units and introduced the words *volt, amp, farad, ohm,* and so on, which have entered into universal usage. The problems of metrology, however, did not end then, and subsequent conventions changed some definitions of the standards. Their practical values changed, too, albeit by small amounts.

In 1873 Sir William met a Miss Blandy, daughter of a wealthy landowner, in Madeira, where he was for a telegraphic enterprise; in 1874 he married her. His new wife liked social life and entertaining. They built a large country house in Largs, near Glasgow, in Scottish baronial style, and as early as 1881 they equipped it with electric light.

Thomson had been interested in politics for years, evolving from very liberal ideas to liberal unionism and imperialism — from Gladstone to Lord Rosebery and Lord Salisbury. In 1892 he was elevated to the peerage in a House of Lords that still had political influence and became Baron Kelvin of Largs. John William Strutt,

Lord Kelvin. (Courtesy AIP Niels Bohr Library.)

Lord Rayleigh (1842 – 1919), was one of the two peers who introduced him to the House of Lords.

Lord Kelvin maintained an unflagging activity to a very advanced age and was interested in any new development in physics. He saw the discoveries of x-ray and radioactivity and befriended the Curies. At the time of his death relativity and quantum theory were already burgeoning. His own physics, however, became dated. In his Baltimore lectures of 1884, published in 1904, admirable under certain aspects, he criticized the electromagnetic theory of light for reasons that would not impress us today. Lord Kelvin was deeply attached to mechanical models, which he treated with great facility and imagination. He was a master of classical mechanics and elasticity, and he pined for a reduction of all physics to mechanical schemes. He was not alone in this. Deeper minds, however, had used models as heuristic tools, but had gone beyond them as necessity required. There lies one of the differences between Kelvin and Maxwell, and the superiority of the latter.

The worldly success of Kelvin was one of the greatest recorded for a scientist. When he died on December 17, 1907, he was mourned by the great of England and by the scientists of the world. He was buried in Westminster Abbey next to Isaac Newton.

The Solid Fortress, Thermodynamics: Conservation of Energy — Mayer, Joule, and Helmholtz

The principle of conservation of energy developed through a long period of widening observations and generalizations until, around the middle of the nineteenth century, it attained a central status in every science, not only in physics. It has remained one of the pillars of modern science ever since, although it had to be further combined with the principle of conservation of mass in the work of Einstein.

In mechanics, it was known by the early nineteenth century that when material points interact by a force derivable from a potential, the sum of the potential and kinetic energies is constant. That theorem hinted at a generalized conservation, including heat, if heat was identified with motion. Most probably, many of the thinkers on the subject were influenced by that consideration. The discoveries in electricity, however, posed new problems on how to fit them into the picture.

The word *energy* was centuries old when it was first used by Thomas Young in 1807 to signify the product mv^2 [not $mv^2/2$], but its use was far from precise. Even such clear thinkers as Carnot and Helmholtz used words such as *puissance motrice*, literally "moving power," and *Erhaltung der Kraft*, literally "conservation of force," when they meant work or energy. Worse, the nomenclature was not constant and, in some cases, downright sloppy. Indefinite formulation and uncertainties in nomenclature make it impossible to attribute the discovery of conservation of energy to one person. Several contributed, but Julius Robert Mayer (1814 – 1878), Joule, and Helmholtz are outstanding. I did not mention Sadi Carnot, who certainly was one of the first to have clear ideas on the subject, because his work was not published until 1878.

Mayer was born in 1814 in Heilbronn, the son of an apothecary. It was the year in which Napoleon was retreating from Russia, and Germany was experiencing the beginning of its nationalistic evolution. Mayer studied medicine at the ancient University of Tübingen. In February 1840 he embarked for one year as a physician on a ship destined for Djakarta, Indonesia. He did not have much to do on board, and he turned to semiphilosophical reflections that ultimately led him to his fundamental new conception of the conservation of energy. His voyage and its consequences may remind us of Darwin's travels, so fateful for the theory of evolution.

In Djakarta Mayer bled a sailor and was surprised by the red color of venous blood, which in other climates was very dark. From this casual observation, he slowly arrived at the idea of the conservation of energy. The first step was the thought that the blood's color was due to the hot climate and the diminished oxidation needed to maintain the body temperature. In that there was a vague hint of the connection of heat to energy. The first published fruit of his meditations is an article of March 1842 in the *Annalen der Chemie*, edited by Justus von Liebig (1803–1873), and entitled "Bermerkungen über die Kräfte der unbelebten Natur" ("Remarks on the Forces of Inanimate Nature"). Previous writings had been refused by the *Annalen der Physik und Chemie*, but the publisher was not too much to blame. Mayer's article contained manifest errors of mechanics and it was anything but clear. Even the article that was published by Liebig is in a philosophical garb that makes it unpalatable to a person not versed in German philosophical style. Nor would his arguments appear very clear and cogent to a modern reader. The redeeming feature is that Mayer gives the mechanical equivalent of the calorie. Unfortunately, he does not say how he reaches that all-important result. From later writings it is clear that what he did was to evaluate the difference between the specific heats at constant pressure and constant volume for a perfect gas. Here is, in modern language, the gist of his reasoning:

Let us measure, in calories, the difference between the molar heat capacity at constant pressure and at constant volume for a gas. C_p is greater than C_V because when we heat the gas it expands and performs work $p_0 \Delta V$, where p_0 is the constant pressure and ΔV the increase in volume. The corresponding energy is supplied to the gas by heat during the heating. If the gas, expanding in a vacuum without making work, does not change temperature, as demonstrated in the Gay-Lussac experiment, we may write

$$C_p - C_V = p_0 \Delta V$$

Experiment gives approximately 1.99 cal/mol·grad. But the gas laws give

$$p_0 V = p_0 V_0 (1 + \alpha\theta) = p_0 V_0 \alpha \left(\frac{1}{\alpha} + \theta \right)$$

and thus, if we change temperature by one degree at constant pressure,

$$\alpha p_0 V_0 = p_0 \Delta V.$$

This quantity may be measured in mechanical units and, with $\alpha = 1/273$, $p_0 = 1$ atm, and $V_0 = 22.4 \; l$, gives $p_0 V_0 = 8.31$ joules/mol grad. (The reader may easily check that $p_0 V_0$ is the universal gas constant usually indicated by R.)

Comparison of the two measurements gives the mechanical equivalent of the calorie. By inserting the numbers he could find in the literature, Mayer obtained for the mechanical equivalent of a calorie: 1 cal = 3.65 kg (the correct value is 2.34).

Mayer wrote two more important publications. He privately printed the book *Organic Motion and Its Relation to Metabolism,* which was an investigation of energy in animal life, with polemic overtones toward Liebig. In 1846 he wrote *Contributions to Celestial Dynamics,* in which he investigated the origin of solar energy, considering the fall of meteorites as a possibility.

Mayer had married in 1842 and settled in Heilbronn. He was somewhat conservative politically, which produced a certain coolness with his liberal brother, especially in the politically explosive years around 1848. In 1850 he attemped suicide by jumping from a window, and from then on, he was intermittently afflicted by insanity, for which he had to be repeatedly hospitalized.

Although the importance of Mayer's work was not appreciated at the time it appeared, it was amply recognized later in his life. Helmholtz, who had read his papers only around 1852, gave him full credit; Clausius and Tyndall corresponded with him; in 1870 he was made a corresponding member of the French Academy, and in 1871 he received the Copley Medal of the Royal Society. He died of tuberculosis in 1878.

I cannot convince myself that Mayer had a great influence on physics. To put it bluntly, and perhaps exaggerating, he was and remained a dilettante. Without him the conservation of energy would have emerged at the same time as it did, and I cannot compare his vague and imprecise conceptions with the subsequent work of Helmholtz and the superb experimental achievements of Joule, who demonstrated the quantitative relations between heat and mechanical work. As a small example of Mayer's style I quote:

> Ex nihilo nil fit (nothing comes from nothing). We call force an object which causes motion when it is applied. What chemistry does for matter, physics must do for force. The sole purpose of physics is to recognize the different forms of force, to investigate the conditions of their metamorphoses. The creation or destruction of forces is beyond the limits of human thought and action.

Years earlier, in 1839, Marc Seguin (1786–1875) had said, "There must be an identity of nature between heat and motion, in such a way that these two phenomena are the manifestation, in a different form, of one and the same cause. These ideas had been transmitted to me by my uncle Montgolfier" (of balloon fame). Ludvig August Colding (1815–1888), the chief engineer of the city of Copenhagen, also had ideas on the conservation of energy and tried to determine the mechanical equivalent of the calorie. They were not the only ones, and much of their work was independent, showing that the idea was in the air. They are all valuable precursors. On the other hand, a person of the stature of Thomson did not grasp the conservation of energy until he was instructed by Joule, and even then it took him some time before he could plumb its implications.

The first to put the new ideas on a solid experimental basis and to give numbers that have stood the test of time was Joule.

James Prescott Joule was born on Christmas Eve in the year 1818 at Salford near Manchester, England, the son of a prosperous brewer. He and his brother were privately taught; among their teachers was John Dalton, then about seventy, of atomic fame. James was of delicate health and slightly deformed. He was attracted to mechanical play, which evolved into serious scientific research. At the beginning of his work, when he was about eighteen years old, he hoped to be able to build perpetual motion machines based on electrical batteries. He soon recognized his error, and the investigation was transformed into an inquiry on the heat developed by electric currents. In 1840 he knew that the heat produced in a wire per unit time was proportional to the square of the current through the wire. He also investigated the case in which the current was not generated by a battery, but by electromagnetic induction. For that purpose, he built a motor generator and measured the work and heat balance very ingeniously and concluded, "We have therefore in magneto electricity an agent capable by simple mechanical means of destroying or generating heat." He then ascertained that there was a fixed rate at which the transformation of heat into work or vice versa occurred.

To understand the problems he faced, one has to read in his papers how difficult it was to measure a current or even to define a proper unit of current. The problems involving all electrical measurements were both technical and conceptual. Joule presented a semipopular but substantial account of his work up to 1847 in a lecture he read at the reading room of St. Ann's Church in Manchester. It was published in the *Manchester Courier* through the good offices of his brother, who was music critic for the newspaper. The following is an excerpt from the article:

> The general rule, then, is, that wherever living force is *apparently* destroyed, whether by percussion, friction, or any similar means, an exact equivalent of heat is restored. The converse of this proposition is also true, namely, that heat cannot be lessened or absorbed without the production of living force, or its equivalent attraction through space. Thus, for instance, in the steam-engine it will be found that the power gained is at the expense of the heat of the fire,—that is, that the heat occasioned by the combustion of the coal would have been greater had a part of it not been absorbed in producing and maintaining the living force of the machinery. It is right, however, to observe that this has not as yet been demonstrated by experiment. But there is no room to doubt that experiment would prove the correctness of what I have said; for I have myself proved that a conversion of heat into living force takes place in the expansion of air, which is analogous to the expansion of steam in the cylinder of the steam-engine. But the most convincing proof of the conversion of heat into living force has been derived from my experiments with the electro-magnetic engine, a machine composed of magnets and bars of iron set in motion by an electrical battery. I have proved by actual experiment that, in exact proportion to the force with which this machine works, heat is abstracted from the electrical battery. You see, therefore, that living force may be converted into heat, and that heat may be converted into living force, or its equivalent attraction through space. All, three, therefore—namely, heat, living

force, and attraction through space (to which I might also add *light,* were it consistent with the scope of the present lecture) — are mutually convertible into one another. In these conversions nothing is ever lost. The same quantity of heat will always be converted into the same quantity of living force. We can therefore express the equivalency in definite language applicable at all times and under all circumstances. Thus the attraction of 817 lb. through the space of one foot is equivalent to, and convertible into, the living force possessed by a body of the same weight of 817 lb. when moving with the velocity of eight feet per second, and this living force is again convertible into the quantity of heat which can increase the temperature of one pound of water by one degree Fahrenheit. The knowledge of the equivalency of heat to mechanical power is of great value in solving a great number of interesting and important questions. In the case of the steam-engine, by ascertaining the quantity of heat produced by the combustion of coal, we can find out how much of it is converted into mechanical power, and thus come to a conclusion how far the steam-engine is susceptible of further improvements. Calculations made upon this principle have shown that at least ten times as much power might be produced as is now obtained by the combustion of coal. Another interesting conclusion is, that the animal frame, though destined to fulfil so many other ends, is as a machine more perfect than the best contrived steam-engine — that is, is capable of more work with the same expenditure of fuel.

Joule tried most ingenious experiments in a variety of ways. The action of paddles in a liquid that we find in all texts is only one of them. He tried compression of gases, inelastic collisions of solids, and others. His experimental technique was excellent. The temperature differences that he had to measure were usually extremely small, but he had mastered the art of such measurements, and his numbers have stood the test of time. I have already mentioned his meeting with Thomson. An important collaboration followed, directed at finding, with greater precision than Gay-Lussac had done, the thermal effect of the expansion of a gas into a vacuum. By then, this could be interpreted as the dependency of the internal energy of the gas on volume.

Joule could reason in a clear and quantitative way, although he was not much of a mathematician. Thus, in 1848 he studied a kinetic model of a gas and found once more Bernoulli's result, explaining the equation of state and, furthermore, the absolute value of the molecular velocity. The following is a short excerpt of his discourse, in which he does not even use formulas:

Let us suppose an envelope of the size and shape of a cubic foot to be filled with hydrogen gas, which, at 60° temperature and 30 inches barometrical pressure, will weigh 36.927 grs. Further, let us suppose the above quantity to be divided into three equal and indefinitely small elastic particles, each weighing 12.309 grs.; and, further, that each of these particles vibrates between opposite sides of the cube, and maintains a uniform velocity except at the instant of impact; it is required to find the velocity at which each particle must move so as to produce the atmospherical pressure of 14,831,712 grs. on each of the square sides of the cube. In the first place, it is known that if a body moving with the velocity of $32\frac{1}{6}$ feet per second be opposed, during one second, by a pressure equal to its weight, its motion will be stopped, and that, if the pressure be continued one second longer, the particle will acquire the velocity of $32\frac{1}{6}$ feet per second in the contrary direction. At this velocity there will be $32\frac{1}{6}$ collisions of a particle of 12.309 grs. against each side of the cubical vessel in every two seconds of

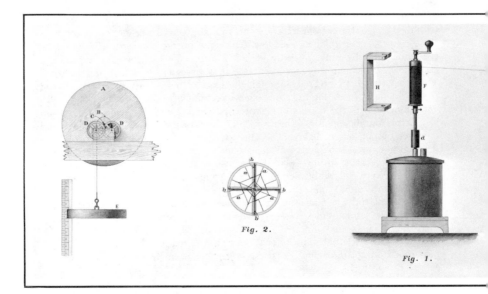

Fig. 2.

Fig. 1.

time; and the pressure occasioned thereby will be $12.309 \times 32\frac{1}{6} = 395.938$ grs. Therefore, since it is manifest that the pressure will be proportional to the square of the velocity of the particles, we shall have for the velocity of the particles requisite to produce the pressure of 14,831,712 grs. on each side of the cubical vessel,

$$v = \sqrt{\left(\frac{14,831,712}{395.938}\right)} \, 32\tfrac{1}{6} = 6225 \text{ feet per second.}$$

The above velocity will be found equal to produce the atmospheric pressure, whether the particles strike each other before they arrive at the sides of the cubical vessel, whether they strike the sides obliquely, and, thirdly, into whatever number of particles the 36.927 grs. of hydrogen are divided.

If only one half the weight of hydrogen, or 18.4635 grs., be enclosed in the cubical vessel, and the velocity of the particles be, as before, 6225 feet per second, the pressure will manifestly be only one half of what it was previously; which shows that the law of Boyle and Mariotte flows naturally from the hypothesis.

The velocity above named is that of hydrogen at the temperature of $60°$; but we know that the pressure of an elastic fluid at $60°$ is to that at $32°$ as 519 is to 491. Therefore the velocity of the particles at $60°$ will be to that at $32°$ as $\sqrt{519} : \sqrt{491}$; which shows that the velocity at the freezing temperature of water is 6055 feet per second.

In the above calculations it is supposed that the particles of hydrogen have no sensible magnitude, otherwise the velocity corresponding to the same pressure would be lessened.

Since the pressure of a gas increases with its temperature in arithmetical progression, and since the pressure is proportional to the square of the velocity of the particles, in other words to their *vis viva,* it follows that the absolute temperature, pressure, and *vis viva* are proportional to one another, and that the zero of temperature of $491°$ below the freezing-point of water. Further, the absolute heat of the gas, or, in other words, its capacity, will be represented by the whole amount of *vis viva* at a given temperature. The specific heat may therefore be determined in the following simple manner: —

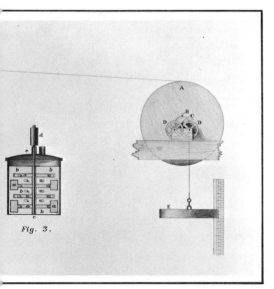

Fig. 3.

Joule's original water-friction experiment for determining the mechanical equivalent of heat. (Crown copyright, Science Museum, London.)

The velocity of the particles of hydrogen, at the temperature of 60°, has been stated to be 6225 feet per second, a velocity equivalent to a fall from the perpendicular height of 602,342 feet. The velocity at 61° will be 6225 $\sqrt{\dfrac{520}{519}} = 6230.93$ feet per second, which is equivalent to a fall of 603,502 feet. The difference between the above falls is 1160 feet, which is therefore the space through which 1 lb. of pressure must operate upon each lb. of hydrogen, in order to elevate its temperature one degree. But our mechanical equivalent of heat shows that 770 feet is the altitude representing the force required to raise the temperature of water one degree; consequently the specific heat of hydrogen will be $\dfrac{1160}{778} = 1.506$, calling that of water unity.

The specific heats of other gases will be easily deduced from that of hydrogen; for the whole *vis viva* and capacity of equal bulks of the various gases will be equal to one another; and the velocity of the particles will be inversely as the square root of the specific gravity. Hence the specific heat will be inversely proportional to the specific gravity, a law which has been arrived at experimentally by De la Rive and Marcet.

[The units used are: 1 grain = 0.0648 gram, 1 foot = 30.48 cm. The degrees are Fahrenheit.]

This paper shows how well Joule reasoned and also his modest mathematical skill, but does not bring out his supreme ability, that of making precise experiments and ferreting out the causes of error with exceptional ingenuity and tenacity. Joule made the excursion into kinetic theory in 1848, when he was still occupied with the experimental determination of the mechanical equivalent of heat, the main goal of his scientific life.

Joule had been elected to the Royal Society in 1850, at the age of thirty-two, and two years later he received its Royal Medal. Public recognition was also given him by many foreign academies. Regrettably, his scientific productivity diminished at an early age, especially as far as ideas are concerned, although he kept at his measurements. In 1875 the British Association entrusted him with the task of making the best determination of the mechanical equivalent of the calorie, and he obtained the value of 4.15, which compares favorably with the present value of 1 cal = 4.184 joules.

Severe economic reverses hit him in 1875, and the rich amateur found himself in financial straits. Friends secured a pension for him of £200 per year, which allowed him to live modestly, but comfortably. At the age of fifty-five, his health deteriorated and his work slowed. His last paper is dated 1878, when he was sixty years old, although he lived to be seventy-one. Two years before his death he told his brother, "I have done two or three little things, but nothing to make a fuss about." I believe most physicists would be happy to have done even one of his little things. His modesty was utterly sincere, and perhaps he would have been surprised (we are not) to find himself commemorated by a tablet in Westminster Abbey and by having his name given to the unit of energy.

The third great contributor to the establishment of the conservation of energy was Hermann von Helmholtz, whose paper "On the Conservation of Force" appeared in 1847, when the author was twenty-six years old. He was a young medical doctor and the paper was his debut in physics. The paper was refused by the

Hermann von Helmholtz (1821–1894) made great discoveries in mathematics, physics, and physiology. This photograph was taken around the time he wrote an epochal paper on the conservation of force (energy). (Courtesy of the Library, University of California, Berkeley.)

editor of the *Annalen der Physik*, Johann Poggendorff (1796–1877), because it was too long, too theoretical, and without enough experimental content. Poggendorff, however, in answering Heinrich Gustav Magnus (1802–1870), who had submitted the manuscript, advised that it be published as a separate pamphlet, and Helmholtz followed the advice.

The work bears some resemblance to Mayer's previous publications, which were unknown to Helmholtz at the time. It is, however, on much more solid ground and contains less philosophy. It gives innumerable examples taken from mechanics, heat, electricity, and chemistry. He gives numbers based on Joule's solid measurements. In evaluating its importance, I shall quote Maxwell:

> To appreciate the full scientific value of Helmholtz's little essay on this subject we should have to ask those to whom we owe the greatest discoveries in thermodynamics and other branches of modern physics, how many times they have read it over, and how often during their researches they felt the weighty statements of Helmholtz acting on their minds like an irresistible driving-power.

We cannot add to the words of such a judge.

Helmholtz was born in Potsdam in 1821, the son of a high-school teacher and the oldest of four brothers. On his seventieth birthday, there was a great ceremony in his honor. On that occasion he gave an autobiographical speech, and the following excerpts tell us an unusual amount of the inner motives and feelings of a great physicist. In the original German, the speech also has considerable literary charm.

> During my first seven years I was a delicate boy, confined for long periods to my room and often to bed; nevertheless, I had a strong inclination toward several occupations and activities. My parents busied themselves a good deal with me, while picture books and games, especially games with wooden blocks, filled the rest of my time. In addition, reading came fairly early, and this, of course, greatly increased the range of my occupations. A defect among my mental powers showed itself, however, almost as early: I had a poor memory for unrelated facts. The first indication of this was, I believe, the difficulty I had in distinguishing between left and right. Later, when I began the study of languages at school, I had greater difficulty than others in learning vocabularies, irregular grammatical forms, and peculiar forms of expression. I could barely master history as it was then taught to us. To learn prose by heart was martyrdom. This defect has, of course, grown and has been a vexation to me in my later years. . . .
>
> The most perfect mnemotechnical method, however, is the knowledge of the laws of phenomena. I first began to learn such laws in geometry. From the time of my childhood playing with wooden blocks, the relations among spatial dimensions were well known to me by actual perception. I knew well, and without much reflection, what sorts of figures would be produced when bodies of regular shape were placed next to one another. When I began the scientific study of geometry, all the facts which I was supposed to learn were perfectly familiar to me, much to my teacher's astonishment. As nearly as I can remember, this occurred at the elementary school attached to the Potsdam Training College, the school I attended up to my eighth year. Rigorous scientific methods, on the other hand, were new to me, and with their help I saw disappear the difficulties which had hindered me in other subjects.

One thing was lacking in geometry: it dealt exclusively with abstract spatial forms, while I delighted in complete reality. As I grew bigger and stronger, I traveled about a good deal in the neighborhood of my native town of Potsdam with my father or my schoolfellows, and I developed a great love of nature. This is perhaps the reason why the first fragments of physics which I learned in the Gymnasium engrossed me much more completely than pure geometric and algebraic studies. . . .

I plunged with pleasure and great zeal into the study of all the books on physics I found in my father's library. They were very old-fashioned: phlogiston still held sway, and galvanism had not grown beyond the voltaic pile. A young friend and I tried, with our limited means, all sorts of experiments about which we had read. The action of acids on our mothers' stores of linens was investigated thoroughly; otherwise we had but little success. Most successful, perhaps, was our construction of optical instruments, using the spectacle glasses that were to be had in Potsdam and a small botanical lens belonging to my father. The limitation of our means during these early studies was valuable in that I was compelled always to vary my plans for experiments in all possible ways, until I got them in a form in which I could carry them out. I must confess that many times while the class was reading Cicero or Virgil, both of whom I found very tedious, I was calculating under the desk the path of light rays in a telescope. Even at that time I discovered some optical laws, not ordinarily found in textbooks, but which I afterward found useful in constructing the ophthalmoscope.

And now I was to go to the university. At that time physics was not considered a profession at which one could make a living. My parents were compelled to be very economical, and my father explained to me that he knew of no way I could study physics other than by taking up the study of medicine in the bargain. . . .

The author speaks of his medical studies and then continues:

Finally, in the last year of my career as a student, I realized that Stahl's theory treated every living body as a *perpetuum mobile*. I was fairly well acquainted with the controversies over the subject of perpetual motion, as I had heard it discussed by my father and by our mathematics teachers during my school days. In addition, while a student at the Friedrich-Wilhelm Institute I was helping in the library, and in my spare moments I looked through the works of Daniel Bernoulli, d'Alembert, and other mathematicians of the last century. I thus came to the questions: What relations must exist among the various natural forces for perpetual motion to be possible, and do these relations actually exist? In my memoir "The Conservation of Force" my aim was merely to provide a critical examination of these questions and to present the facts for the benefit of physiologists.

I was quite prepared for the experts to say simply, "We know all that. What is this young doctor thinking about who considers himself called upon to explain it all to us so fully?" To my astonishment, however, the authorities on physics with whom I came in contact received it quite differently. They were inclined to deny the correctness of the law and, because of the heated fight in which they were engaged against Hegel's philosophy of nature, to treat my essay as a fantastic piece of speculation. Only the mathematician Jacobi recognized the connection of my line of thought with that of the mathematicians of the preceding century, defended my investigations, and protected me from misconception. I also met with enthusiastic applause and practical help from my younger friends, especially Emil du Bois-Reymond. They soon brought the members of the newest physical association of Berlin over to my side. I knew little at that time about Joule's researches on the subject and nothing at all about those of Robert Mayer. . . .

Helmholtz tells how some papers on fermentation secured for him the chair of general pathology and physiology at Königsberg, where he invented the opthalmoscope, and continues:

The construction of the ophthalmoscope had a most decisive effect on my position in the eyes of the world. From that time on I met with immediate recognition from the authorities and my colleagues, and with an eagerness to satisfy my wishes. Thus I was able to follow far more freely the impulses of my desire for knowledge. I must, however, say that I attribute my success in great measure to the fact that, possessing some geometric understanding and equipped with a knowledge of physics, I had the good fortune to be thrown into medicine, where I found in physiology a virgin territory of great fertility. Furthermore, I was led by my knowledge of vital processes to questions and points of view which are usually foreign to pure mathematicians and physicists.

Up to that time I had only been able to compare my mathematical ability with that of my fellow students and medical colleagues; that I was for the most part superior to them in this respect did not, perhaps, say very much. Moreover, mathematics was always regarded at school as a subject of secondary importance. In Latin composition, on the other hand, which then determined the palm of victory, half of my fellow students were ahead of me.

In my own mind my researches were simple logical applications of the experimental and mathematical methods which had been developed in the sciences and which, by slight modifications, could easily be adapted to the particular problems at hand. Colleagues and friends who, like myself, devoted themselves to the physical aspect of physiology made discoveries no less surprising.

In the course of time, however, matters could not remain at that stage. Problems which could be solved by established methods had gradually to be handed over to the students in my laboratory, and I had to turn to more difficult researches, where success was uncertain, where standard methods left the investigator stranded, or where the method itself had to be worked out.

In these regions closer to the boundaries of our knowledge I have also succeeded in many things experimental and mathematical — I do not know if I may add philosophical. With respect to the first, like anyone who has attacked many experimental problems, I had come to be a person of experience, acquainted with many plans and devices, and my youthful habit of considering things geometrically had developed into a kind of mechanical intuition. I felt, intuitively as it were, how stresses and strains were distributed in any mechanical arrangement. This is a faculty also met with in experienced mechanics and machinists; I had the advantage over them, however, in that I was able to make especially important and complicated relations clear by means of theoretical analysis.

I have also been in a position to solve several problems in mathematical physics, some of which the great mathematicians since the time of Euler had worked on in vain — for example, problems concerning vortex motion and the discontinuity of motion in fluids, the problem of the motion of sound waves at the open ends of organ pipes, and so on. But the pride which I might have felt about the final result of these investigations was considerably lessened by my knowledge that I had only succeeded in solving such problems, after many erroneous attempts, by the gradual generalization of favorable examples and by a series of fortunate guesses. I would compare myself to a mountain climber who, not knowing the way, ascends slowly and toilsomely and is often compelled to retrace his steps because his progress is blocked; who, sometimes by reasoning and sometimes by accident, hits upon signs of a fresh path, which leads him a little farther; and who finally, when he has reached his goal, discovers to his annoyance a royal road on which he might have ridden up if he had been clever enough to find the right starting point at the beginning. . . .

As I have often found myself in the unpleasant position of having to wait for useful ideas, I have had some experience as to when and where they come to me which may perhaps be useful to others. They often steal into one's train of thought without their significance being at first understood; afterward some accidental circumstance shows how and under what conditions they originated. Sometimes they are present without our knowing whence they came. In other cases they occur suddenly, without effort, like an inspiration. As far as my experience goes, they never come to a tired brain or at the desk.

I have always had to turn my problems about in my mind in all directions, so that I could see their turns and complications and think them through freely without writing them down. To reach that stage, however, was usually not possible without long preliminary work. Then, after the fatigue of the work had passed away, an hour of perfect bodily repose and quiet comfort was necessary before the fruitful ideas came. Often they came in the morning upon waking, . . . as Gauss also noted. But, as I once stated at Heidelberg, they were most apt to come when I was leisurely climbing about on wooded hills in sunny weather. The slightest quantity of alcohol seemed to frighten them away. . . .

I have also entered another region to which I was led by investigations of sensation and sense perception, namely, the theory of knowledge. Just as the physicist must examine the telescope and galvanometer with which he is working in order to get a clear conception of what he can attain with them and of how they may deceive him, so it seemed to me necessary to investigate our powers of thought. Here also we are concerned only with a series of factual questions to which definite answers can and must be given. We have specific sense impressions, as a consequence of which we know how to act. The observable results of actions usually agree with what was anticipated; sometimes, however, as in cases of subjective impressions, they do not. These are all objective facts, and it is possible to find the lawful relations among them. My principal conclusions were that sensory impressions are only signs of the constitution of the external world and that the interpretation of these signs must be learned by experience. . . .

It has been my aim to explain to you how the history of my scientific endeavors and achievements, as far as they go, appears when seen from my own point of view. Perhaps you will now understand why I am surprised at the unusual amount of praise you have heaped upon me. With respect to my own estimate of myself, my achievements have had primarily the following value: they have provided a measure of what I must still attempt. They have not, I hope, led me to self-admiration. I have often enough seen how injurious for a scholar an exaggerated sense of self-importance may be, and I have always known that a rigorous self-criticism of my own work and of my own capabilities is the protection and palladium against this fate. It is really only necessary, however, to keep one's eyes open for what others can do and for what one cannot do, to avert that great danger. Moreover, as regards my own work, I do not believe that I have ever finished correcting the last proof of a memoir without finding, in the course of the next twenty-four hours, a few points which I could have done better or more carefully. . . .

During the first half of my life, when I still had to work to establish myself, I would not say that higher ethical motives were not present, along with a desire for knowledge and a feeling of duty as a servant of the state. It was difficult, however, to be certain of their presence as long as egoistic motives to work were still there. This is perhaps the case with most investigators. Afterward, however, when an assured position has been attained and those who have no inner impulse toward science may cease their labors, a higher conception of one's relation to humanity does influence those who continue to work. They gradually learn from their own experience how the thoughts which they have expressed, either through their writings or through oral instruction, continue to act on other men and possess, as it were, an independent life.

[*Selected Writings of Hermann von Helmholtz,* Russell Kahl, editor and translator (Middletown, CT: Wesleyan University Press: 1971), pp. 466–477. © 1971 by Russell Kahl; reprinted by permission of Wesleyan University Press.]

Helmholtz's speech provides remarkable insight into the psychology of a great physicist. It is written with apparent candor by a man who has a great career behind him, and who, at the time (1885), held a position in Germany that was described by Michael Pupin (1858–1935), then a student in Berlin, as, "Next to

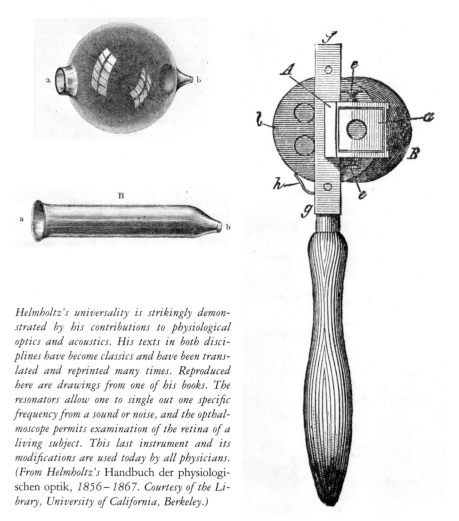

Helmholtz's universality is strikingly demonstrated by his contributions to physiological optics and acoustics. His texts in both disciplines have become classics and have been translated and reprinted many times. Reproduced here are drawings from one of his books. The resonators allow one to single out one specific frequency from a sound or noise, and the opthalmoscope permits examination of the retina of a living subject. This last instrument and its modifications are used today by all physicians. (From Helmholtz's Handbuch der physiologischen optik, 1856–1867. Courtesy of the Library, University of California, Berkeley.)

Bismarck and the old Emperor, he was at that time the most illustrious man in the German Empire."

To Helmholtz's description of his career, we can add a number of facts. In 1855 he moved to Bonn as professor of physiology and anatomy. In the meantime, he had visited England in 1853 and become friendly with a number of scientists in that country. Describing his travels to his wife, he called "Berlin . . . a village compared to London; in size and cultural activities." By 1858 he was already famous and moved to Heidelberg. There he did original research on vision and audition, writing two large treatises on the sensation of sound and on physiological optics. Those classic books exerted enormous influence on the whole field of physiology of the senses.

In 1870 he became professor of physics at Berlin, with a huge salary of 4,000 talers per year, about four times that of a normal university professor. There he studied mathematical hydrodynamics, obtaining fundamental results on vortex theory. He also initiated a critical examination of the current electrical theories, comparing those of Wilhelm Weber, Carl Neumann, and others with those of Maxwell. He tried to assess the self-consistency of all those theories and their compatibility with the conservation of energy, to see in what respect they made different predictions, and, finally, to see how those predictions could be tested by experiment. That important program was an inspiration to Hertz, who ultimately brought it to fruition.

The formal lectures of Helmholtz, on the other hand, were not good. He did not prepare sufficiently, and he was a poor improviser. The result was that at Berlin, the greatest center of German physics, the two foremost professors were mediocre lecturers. Gustav Robert Kirchhoff (1824–1887) put the audience to sleep, and Helmholtz was hardly intelligible!

In 1885 Helmholtz became director of the newly created Reichsanstalt, an institution first financed by the munificence of von Siemens, the electrical industrialist, who contributed 50,000 marks to it. It was both a research institution and a bureau of standards. In the research part, both pure and applied research were contemplated. The institution also had patriotic overtones, as a symbol of unity for the newly founded German empire. Helmholtz devoted himself to the establishment of the Reichsanstalt with his usual energy and intelligence, but he also preserved a connection to the university, where he occasionally lectured.

With the passage of time, his interests had become more theoretical. He tried to give a mechanical foundation to thermodynamics and to formulate a minimum principle, generalizing the least-action principle, hoping to find a foundation of great generality for all of physics. The mathematician Leopold Kronecker (1823–1891), in congratulating him on his sixty-seventh birthday, expressed the hope that he would become a pure mathematician. In fact, some of his work connecting his physiological studies to studies on the foundations of mathematics may have pointed in that direction.

Helmholtz was remarkable also for his students. People from many countries worked in his institute. Some of them became leaders in their own right—

Heinrich Rudolf Hertz (1857–1894), Henry Augustus Rowland (1848–1901), Albert Abraham Michelson (1852–1931), and Michael Pupin (1858–1935) are among the more famous. The number of the less famous is also great, and they transferred some of Helmholtz's spirit to many countries. In many ways he was the German counterpart to his friend Lord Kelvin, even in terms of official distinction: The Kaiser made Helmholtz "von" and gave him the title "Geheimrath mit dem Praedikat Exzellenz."

Helmholtz also must have had exceptional personal qualities, but they do not clearly emerge from the honey of his official biography. However, the letters from Hertz and the testimony of Max Planck (1858–1947) and of Pupin (certainly very different persons) all attest to a considerable measure of personal affection, in addition to infinite respect.

Planck, who became professor of theoretical physics at Berlin in 1889, says.

> I came to know him [Helmholtz] also from his human side and to appreciate him for it as much as I had always done for his science. In the entirety of his personality, in his unerring judgment, and his unassuming manners he embodied the dignity and truthfulness of his science. A human goodness that touched me deeply accompanied all this. When he looked at me in conversations, with his calm, inquisitive and penetrating eyes, that however were fundamentally kind, I was overcome by a feeling of complete childish trust. I could have confided to him without any reserve all that was on my mind, sure that I would have found in him a just and benevolent judge. A word of recognition or of praise from his mouth could make me happier than any external success.

All this does not contradict the impression of others that in Helmholtz's institute, there was strict discipline and a certain amount of stiffness that was characteristic of Prussia at that time. Ludwig Boltzmann (1844–1906) clearly had mixed feelings on the subject, and while he said that Helmholtz was the only one who understood and appreciated him (a great exaggeration), he changed his mind about accepting a Berlin position after a dinner with Frau Helmholtz. She had told Boltzmann candidly that Berlin was not for him, possibly rendering him a great service.

Thermodynamics Perfected: Rudolf Clausius

Carnot had created thermodynamics by initiating its powerful methods and discovering its second principle. Mayer, Joule, and Helmholtz had formulated the first principle. The synthesis of both is due to Clausius and, later, to Thomson. Clausius is much less known among nonphysicists than Kelvin or Helmholtz, but he has a truly important place in twentieth-century physics as one of the founders of thermodynamics and of kinetic theory. His pivotal position in thermodynamics was well expressed by Josiah Willard Gibbs (1839–1903), the last and most perfect epitome of classical thermodynamics and statistical mechanics: "If we say in the words used by Maxwell some years ago, that 'thermodynamics is a science with secure foundations, clear definitions, and distinct boundaries,' and ask when those foundations were laid, those definitions fixed, and those boundaries traced, there can be but one answer. Certainly not before the publication of that memoir." He refers to a memoir by Clausius "On the Motive Power of Heat, and on the Laws which can be deduced from it for the Theory of Heat," which appeared in 1850 in volume 79 of Poggendorff's *Annalen der Physik.*

To the qualifications given by Maxwell to thermodynamics, we add that in the subsequent revolutions of twentieth-century physics, it stood firm as a rock and that the great innovators, such as Planck and Einstein, used it as their anchor when everything seemed to be put in doubt.

Clausius was the first to see clearly through the apparent contradictions of the then-prevalent theories of heat and to give a systematic treatment of the new science. His memoirs and, later, his treatise gave thermodynamics the form under which it is still taught today.

Rudolf Clausius was born at Köslin, in Prussia, in 1822, the son of a school inspector and Protestant pastor, who was also the principal of a small school that the son attended. The son went to Stettin for his high-school education and later to the University of Berlin, where he was attracted by historical studies first, but soon he passed to scientific studies. Georg Simon Ohm (1789–1854), the discoverer of Ohm's law, and the mathematicians Gustav Peter Dirichlet (1805–1859) and Jakob Steiner (1796–1863), were among his professors. He had started at the university in 1840, but he could not finish for financial reasons, and from 1843 to

Rudolf Clausius (1822–1888), professor at several German universities and finally at Bonn, was one of the founders of thermodynamics. He reconciled the ideas of Carnot with conservation of energy and introduced the concept of entropy as a measure of the reversibility of a phenomenon. He also made fundamental contributions to the kinetic theory of gases. (Deutsches Museum, Munich.)

1850, he taught in a Berlin gymnasium, while continuing his studies at the university, where he received his doctorate in 1848. He then wrote some papers on optical phenomena in the atmosphere and took a position at the Royal Artillery School; he also became Privat Dozent at the University of Berlin. In 1850 he presented his first capital memoir to the Prussian Academy of Sciences, on what we now call thermodynamics. That was the work in which he reconciled Carnot with the conservation of energy; Clausius was then twenty-eight years old. The next year he met Tyndall, the successor to Faraday at the Royal Institution and a gifted popularizer of science, as well as a scientist. Clausius struck a solid friendship with him, and Tyndall was godfather to Clausius's oldest child, as well as translator of some of his work and defender in priority claims that filled the scientific journals for some time.

A few years later, Clausius was called to the University of Zurich, and shortly thereafter to the Polytechnicum in the same city. For two years he was in Würzburg, and in 1869 he settled finally in Bonn, where he remained for the rest of his life. In 1859 he had married A. Himpan, with whom he had an especially happy union. Unfortunately, she died in giving birth to their sixth child in 1875. That was a shattering blow for Clausius, who was left alone with the small children. He took care of them with great solicitude, but that may have interfered with his work, which slowed down appreciably. In addition to his family tragedies, Clausius had been seriously wounded in a knee in 1870 during the Franco-Prussian War, where he had served as an ambulance driver. He remained slightly incapacitated for life. In 1886, when his youngest child was eleven years old, he remarried. The bride, S. Sack, was much younger than he, but the marriage was also very successful, even though it lasted only a short time, because in 1888 Clausius contracted pernicious anemia and died. Clausius had been amply recognized with the usual academic honors in his native Germany and in the rest of Europe. In the later years of his life, he wrote on electrical theory, too, but his fame rests on his work in thermodynamics and kinetic theory of gases.

In systematizing thermodynamics, Clausius expressed the first principle of thermodynamics by the equation

$$dQ = dU + dW$$

where U is the internal energy of the system, which is the same as the kinetic and potential energy of its molecules. The internal energy is a function of the macroscopic state variables, for instance, p and V. The differential dW is the external work done in supplying the heat. In the case of a gas or a fluid, $dW = p\,dV$. In that equation, the quantity of heat is equated to energy. It is thus implied that there is a mechanical equivalent of heat, the major result of Joule's work. In Clausius's words, the first principle is that "work may transform itself into heat, and heat conversely into work, the quantity of one bearing always a fixed proportion to that of the other."

For the second principle, the situation is more complicated. There, reversibility is the essential concept, and at first I shall consider only reversible processes. Let us now bring our gas or any other fluid from one state to another reversibly. In a p,V plane that operation can be described by a line. Note that for an irreversible transformation that is not possible, because, for example, the pressure may not be the same everywhere in the system.

Let us now perform a Carnot cycle on the fluid as described on page 196. The cycle is composed of an isothermal expansion at temperature T_1 (defined by an arbitrary thermometer), in which the amount of heat Q_1 is supplied. This expansion is followed by an adiabatic expansion, in which the temperature T_2 is reached, on an arbitrary scale. The third step is an isothermal compression at temperature T_2, in which the amount of heat Q_2 is surrendered by the system. The fourth step, an adiabatic compression, closes the cycle returning the system to the initial temperature T_1. The cycle is *reversible,* and, if *"heat cannot of itself pass from a colder to a hotter body,"* we have, by a reasoning similar to that given on page 205, the relation $Q_1/T_1 = Q_2/T_2$. That relation defines the ratio of the temperatures T_1/T_2, which is independent of the substance used in the cycle and is thus absolute. The above words in italics are Clausius's form of the second principle of thermodynamics. The clause "of itself" is all important. Crudely, it means that the *only* phenomenon occurring is the heat passage, without any compensatory effect. To explain that point in depth would require too much space, and the reader is referred to any book on thermodynamics.

The next step is the generalization of the relation $Q/T_1 = Q_2/T_2$ to arbitrary transformations. One must give the proper sign to the heat supplied and to the heat surrendered, and, furthermore, decompose any arbitrary reversible cycle into an infinity of Carnot cycles. One then obtains the equation

$$\oint \frac{dQ}{T} = 0$$

where the circle on the integral means that the equation obtains for any reversible cycle. It follows that starting from a point A in a p, V plane and going to a point B in the same plane, the integral does not depend on the path followed.

We can thus define a function of the state S_B (called "entropy" by Clausius) as

$$S_B - S_A = \int_A^B \frac{dQ}{T}$$

where the passage from A to B is accomplished reversibly. Since the initial reference state S_A is arbitrary, the entropy is defined except for an arbitrary constant. That constant may be determined by further considerations that go beyond classical thermodynamics and are founded on quantum theory. The Nernst theorem, developed in the early years of the twentieth century, says that "the entropy of every system at absolute zero can always be taken as zero." Of course, it is not necessary to use p and V as variables. Any two variables that define the state, such as T and V, are acceptable.

The equation that defines the entropy may be written as

$$dS = \frac{dQ}{T},$$

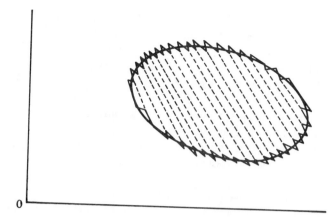

The decomposition of an arbitrary reversible cycle into a collection of Carnot cycles, as introduced by Clausius in The Mechanical Theory of Heat, 1865–1867. *(Drawing from Emilio Segrè,* From X-Rays to Quarks, *W. H. Freeman and Company,* © 1980.)

which, combined with the first principle, also gives the formula

$$dS = \frac{dU + p\,dV}{T}.$$

For an irreversible process, the increase in entropy is larger than $\int_A^B dQ/T$:

$$S_B - S_A \geq \int_A^B \frac{dQ}{T}.$$

For a completely isolated system, no heat is supplied, and $S(B) \geq S(A)$.

Clausius coined the word *entropy* from the Greek τροπή ("transformation") and introduced it in 1865. He had already developed the concept in 1851. The introduction of entropy allows us to use thermodynamics in a semiautomatic way. It does not add conceptually to Carnot's fundamental ideas based on cycles, but it permits us to calculate simply and rapidly, without laborious reasonings based on cycles. On the other hand, the more cumbersome method adheres more perspicuously to experimental facts, although idealized.

The interpretation of entropy, its deep meaning, became one of the central problems for Clausius's contemporaries. In particular, classical mechanics considers only reversible phenomena, while in thermodynamics, entropy grows with time in an irreversible way. The reconciliation of these facts posed a dilemma that was solved in subsequent years by the introduction of probability and statistics into physics, as we shall see.

Clausius expressed the fundamental empirical facts on which thermodynamics is based by the axioms mentioned above. William Thomson chose two other axioms. For the first law, he stated, "When equal quantities of mechanical effects are produced by any means whatever from purely thermal sources, or lost in purely thermal effects, equal quantities of heat are put out of existence or are generated." He ascribed the proof of this fact to Joule. The second axiom he took is, "It is impossible, by means of inanimate material agency, to derive mechanical effect from any portion of matter by cooling it below the temperature of the coldest of the surrounding objects." Thomson and Clausius's axioms are equivalent.

The second principle poses limitations to the transformations of energy. In a homogeneous system all at the same temperature, the energy cannot be transformed into any other form. That circumstance gave rise to speculation on a thermal death of the universe.

Clausius also formulated the two principles into two impressive phrases that are, however, open to objections. First law: "The energy of the universe is constant." Second law: "The entropy of the universe tends to a maximum."

Innumerable applications followed the establishment of the principles of thermodynamics — to imperfect gases, to electrical phenomena, to electrochemistry, to changes of state, to vapors, to capillarity, to electricity, and to more. There is no field of physics to which thermodynamics cannot be applied. That intensive work was accomplished in the latter part of the nineteenth century and still continues. Possibly the most important applications were to chemistry and industrial chemistry. Empiricism was, to an extent, supplanted by a scientific understanding of chemical equilibrium and of what could be accomplished and what could not. Several of the great syntheses, for example, ammonia and nitric acid, were led forward by the science of thermodynamics.

Chapter 6

Kinetic Theory: The Beginning of the Unraveling of the Structure of Matter

Unfortunate Precursors

Thermodynamics is a branch of physics that can be formulated axiomatically, like mathematics, and gives conclusions as sure as its premises without enlightening us on the details or models of the phenomena it considers. Those features are attractive to certain minds, but many physicists have different instincts and have been striving for a long period for a more intuitive foundation.

The behavior of gases has played a major part in the development of our conceptions on matter and on heat. We have to look at the pertinent ideas to appreciate the other great branch of physics, statistical physics, that developed in the nineteenth century.

A gas was often considered as a continuous fluid, and the fundamental concept of pressure is perfectly suited to such a view. Pumps, barometers, and similar instruments do not throw any light on the internal constitution of a gas. When Robert Boyle (1627–1691) and his predecessors and successors worked out "the spring of air," they did not know to what it was due. Speculation was natural, and semistatic models based on springs seemed plausible. On the other hand, the idea of a gas being composed of very small particles, flying in a disordered fashion, perhaps attracting or repelling each other and bouncing on the walls, also had its supporters. Naturally, nobody had any inkling of the size and mass of these particles.

One of the first important contributions to the understanding of the mechanism of pressure was provided by Daniel Bernoulli (1700–1782) in 1738.

Bernoulli could prove by a simple argument that the pressure of a gas obeys the equation,

$$pV = \tfrac{1}{3}Nm\langle v^2 \rangle,$$

where p is the pressure, V the volume, N the number of molecules, mass is m, and $\langle v^2 \rangle$ the mean square velocity. He could not account for the relation of the velocity of the molecules to the temperature of a gas. Later, at the beginning of the nineteenth

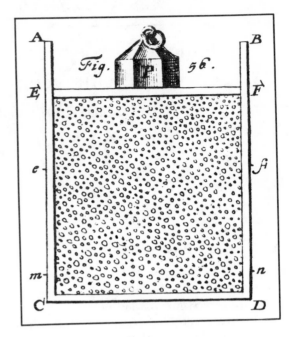

An illustration from Hydrodyna-
mica *(1738) by Daniel Bernoulli
(1700–1782) in support of his
calculation of the pressure of a
gas. Note that, in a real gas at
atmospheric pressure and room
temperature, the average distance
between molecules is roughly 10
times the molecular diameter.
(Courtesy of the Library, University of California, Berkeley.)*

century, when the expansion coefficients of gases were measured and found to be all
equal, the empirical concept of "absolute" temperature [$T = 273 + t$ (centigrade)]
emerged, and it became possible to connect it to the molecular velocity, although
that occurred in various steps and included notable errors.

Another fundamental ingredient to the kinetic theory was provided by
Amedeo Avogadro, Conte di Quaregna (1776–1856). In his long life, he wrote
many books that are now forgotten, but he made one capital contribution that is
fundamental to chemistry and physics. In 1811 he stated the principle that equal
volumes of gases at the same temperature and pressure contain the same number of
molecules. With that rule, Bernoulli's equation, and the absolute scale of temperature, one can write,

$$pV = \tfrac{1}{3}Am\langle v^2 \rangle = RT,$$

where A is the number of molecules in a mole and V is the volume of a mole of gas
(22.4 liters at $0°C$ and 1 atmosphere pressure). It follows that

$$\frac{3}{2}kT = \frac{m\langle v^2 \rangle}{2}$$

with $k = A/R$ a universal constant. The average kinetic energy of a molecule is
proportional to the absolute temperature, and the constant of proportionality is
universal! All this can be found in Joule's paper (see pages 217–218).

The ingredients for some of the foundations of kinetic theory were thus ready by 1811, but they were not put together for many years.

Two of the early investigators in kinetic theory, John Herapath (1790–1868) and John James Waterston (1811–1883), had the right ideas but particularly bad luck in communicating them to the scientific community.

Herapath was born in Bristol, England, the son of a maltster. He showed early mathematical ability, although he worked in his father's business. In 1815 he married and abandoned his previous occupation, setting up a school of mathematics. He wrote his papers on kinetic theory around 1820, but did not succeed in having them accepted for publication by the Royal Society. He published an important paper in 1821 in the *Annals of Philosophy,* but it contained a serious error because it set the temperature proportional to the velocity of the molecules rather than to its square. Later his interest turned to railways and he published a journal devoted to them and wrote extensively for it. He had a difficult personality and he is remembered mainly because he had some influence on his successors; James Prescott Joule, James Clerk Maxwell, and others quoted him in their pioneer work.

Waterston was the son of a sealing wax manufacturer of Edinburgh and he was related to Robert Sandeman, who founded the Sandemanian religious sect followed by Michael Faraday. He grew up in comfortable circumstances and studied at Edinburgh University. He was in London working for the railways when, in 1839, he was offered a teaching job in India. In Bombay he found the leisure he desired, and he used it for scientific studies, which were published but ignored. In 1845 he submitted a truly important paper on kinetic theory to the Royal Society, but it was refused. Extracts had been published by the British Association, but the importance of his results was not understood. In particular, an abstract dated from Calcutta in 1851 and sent to the British Association, if understood, should have alerted the physicists. In 1857 he returned to England and lived there in delicate health. In 1891, Lord Rayleigh (1842–1919), then secretary of the Royal Society, puzzled by a reference he found in a paper on sound by Waterston, retrieved his manuscript from archives and published it in the *Philosophical Transactions* of 1892. Waterston had found the equipartition principle and the absolute value of the mean molecular velocity. If his paper had been known in 1845, it certainly would have influenced the development of the subject. Understandably, Waterston remained rather bitter and contemptuous of learned societies.

The papers of Herapath and Waterston had been collecting dust in the archives of the Royal Society for about ten years when August Karl Krönig (1822–1879) published essentially the same results. Krönig, however, had much more success in calling attention to the kinetic theory. He was a well-established professor at the Realschule (polytechnic school) in Berlin; furthermore, he was the editor of the *Fortschritte der Physik,* an annual survey of physics progress that was well known in the profession. He published his paper in Poggendorff's *Annalen der Physik,* one of the leading physics journals. His main results were the kinetic explanation of Boyle's law and Gay-Lussac's law and other topics anticipated by Waterston.

Again, Clausius and Maxwell

A new departure in kinetic theory was made by Rudolf Clausius (1822–1888) in a paper of 1857, when he had finished his most important thermodynamic work. The title of the paper was "On the Kind of Motion that We Call Heat." He repeats, in part, known things, but with far greater clarity than in the past and with attention to many obscure points. For example, if molecules are complex particles that may have internal vibrations of their component atoms and also rotations of the molecule as a whole, how do these motions contribute to pressure and specific heat?

In a paper written the following year, Clausius answers an objection of the Dutch meteorologist Christoph Hendrik Buys-Ballot (1817–1890), better known for his studies on the circulation of the atmosphere. Buys-Ballot had asked: if the velocity of the molecules is as high as calculated by kinetic theory, how does it happen that if hydrogen sulfide is mixed in one corner of a room, several minutes elapse before it is smelled in another corner? Or, how does tobacco smoke remain so long in immovable layers? In order to answer these objections, Clausius pointed to the mechanism of diffusion in a gas, and, in so doing, had to introduce the concepts of sphere of action, collision cross section σ, and mean free path l.

The sphere of action is defined by considering two molecules, assumed to be little spheres of radius r. If the center of the two molecules come closer than a certain distance $s = 2r$, they collide and are deflected from their trajectories. We express these facts by saying that each molecule is surrounded by a sphere of action of radius s. The collision cross section σ is πs^2, and the mean free path l is, apart from a numerical factor of order one,

$$l = 1/n\sigma,$$

where n is the number of molecules per unit volume.

Clausius noted that at the time nobody had any idea of the magnitude of s, except that it was very small compared to a centimeter. A determination of the mean free path would thus give the handle for finding the absolute size of a molecule and, hence, (assuming that in a solid the molecules touch each other) their number per unit volume, their mass, and similar quantities. The importance of these data is obvious.

In all those investigations, Clausius used mean values of the square of the velocity and similar averages. He never inquired about the velocity distribution, which, nevertheless, influences the results. However, Clausius' memoirs inaugurated a more modern kinetic theory, and from then on, it was unnecessary to read previous work.

Among Clausius's readers was Maxwell, who entered the field with a bang. In a memoir first read at the meeting of the British Association in September of 1859 he gave the velocity distribution of the molecules in a gas (see Appendix 10 and page 168). He also calculated the viscosity coefficient of a gas in terms of molecular quantities. The mean free path enters the equations and thus becomes accessible to experiment. The memoir is divided into three parts, one of which contains errors that were pointed out by Clausius and admitted by Maxwell. But the other two make

Ludwig Boltzmann (1844–1906). (Courtesy of D. Flamm.)

gigantic strides over what was previously known. In the final part, Maxwell finds once more that the kinetic energy per degree of freedom is the same for all particles in thermal equilibrium. This is the equipartition principle, one of the mainstays of classical statistical mechanics. In due course, its violation would become one of the stumbling blocks of classical physics. With those theoretical underpinnings, the measurement of the gas viscosity became a prime experiment, and we have already seen how Maxwell undertook it with the help of his wife.

Ludwig Boltzmann

Other problems of a fundamental nature loomed. Most investigators of kinetic theory had also been active in thermodynamics, and it was clear that there was a connection between the two sciences. There, however, appeared a fundamental difficulty. At the center of thermodynamics stands the fact that the entropy of an isolated system increases with time until it reaches a maximum at equilibrium. The empirical foundation of this fact is the impossibility of a *perpetuum mobile* of the second kind or the arrow of time, as indicated by heat conduction. On the other hand, all mechanical phenomena are time-reversible — that is, if at a given moment we reverse all the velocities of a system of points, the motion is reversed, as if we saw a movie run backward. Is it possible to explain from pure mechanics the irreversibility of natural phenomena? Is the increase of entropy derivable from mechanics? Those fundamental questions worried many of the best minds of the second half of the nineteenth century.

It was natural to try to reduce such a fundamental law as the second principle of thermodynamics to mechanics, and a first attempt was made by Ludwig Boltzmann. He was an Austrian, son of a tax collector, born in 1844 in the splendor of Vienna, soon to be ruled by Franz Joseph. Already by 1848 the nationality problem (which brought the demise of the Empire at the end of World War I) had nearly broken the Hapsburg Empire, but Franz Joseph managed to hold it together and made the cracks barely visible. The great musical tradition of Vienna was carried on by Bruckner, Brahms, Mahler, and the Strauss dynasty. In that atmosphere science and medicine thrived. Above all, Vienna had a civilized way of life that invited comparison with Paris. Vienna thought of itself as the Paris of eastern Europe, and the rest of Europe concurred.

Boltzmann passed his early years in Wels and Linz, small provincial cities, but attended the University at Vienna, where he had a professor, Josef Stefan (1835–1893), who was to become famous for his studies on radiation, and an older friend, Johann Joseph Loschmidt (1821–1895), who was the first to give some reliable numbers concerning molecular magnitudes.

By 1864 Boltzmann had finished his formal study; in 1867 he became Privat Dozent. In 1869 he had his first professorial appointment, in Graz, for mathematical physics, but his stay there was short. He spent some time in 1869 and 1871 in Heidelberg and in Berlin with Gustav Robert Kirchhoff and Hermann von Helmholtz. From 1873 to 1876, he taught mathematics in Vienna, but he then returned to Graz as a professor of experimental physics. He was by now well known, and smart young physicists came to Graz to study with him; prominent among them were Walther Hermann Nernst and Svante August Arrhenius, future leaders in physicochemistry. In 1890, Boltzmann moved to Munich, a much bigger place than Graz, but this was not his last move. In 1894 he succeeded Stefan at Vienna. Then for two years he went to Leipzig, and, finally, he returned to Vienna, where he remained until his death. Although at heart a theoretical physicist, Boltzmann had performed important measurements on the dielectric constant and refractive index of gases and solids to confirm Maxwell's theory (1874). He was an able experimenter, but was handicapped by his severe nearsightedness. In later years his physical health deteriorated. He suffered from severe asthma and headaches, and, at the same time, his eyesight weakened to the point where he was forced to employ a reader. He had attempted suicide while in Leipzig, but he survived. On September 5, 1906, while on vacation in Duino, near Trieste, he killed himself.

The continual changing of residences tells us something of Boltzmann. He was subject to moods that could deepen into true depressions. Although he was recognized as one of the major physicists of his time, he occasionally felt isolated and intellectually abandoned, feelings without true justification. He had received his full share of honorary degrees, including one from Oxford; he was a member of the main academies; and, for his sixtieth birthday, he was celebrated with a festschrift containing contributions by such physicists as Arrhenius, Jacobus Henricus van't Hoff, Lorentz, Ernst Mach, Nernst, Max Planck, Arnold Sommerfeld, Johann Diderik van der Waals, Wilhelm Wien, and many others from England, France, America, Russia, Japan, Italy, and so on.

Boltzmann was very sensitive to intellectual attacks, but that did not prevent him from taking part in bitter polemics, nor did it suggest to him that he treat his adversaries gently. He unmercifully exposed Wilhelm Ostwald (1853–1932), a renowned chemist who for a time was an adversary of atomism and advocated an ill-founded doctrine of "energetics." However, by 1905 Albert Einstein had published his theory of Brownian motion, which, together with preceding experiments on radioactivity, brought direct evidence for the atomistic structure of matter. By 1910, after Boltzmann's death, his main opponents, including Ostwald and Mach, were either converted or very much on the defensive. In any case, even in scientific disagreement, his opponents had respected him.

Boltzmann had an artistic temperament and wrote unusually well. He loved Schiller best among the German poets, to such an extent that he said that without Schiller he could not have existed in a spiritual sense. He was also a good pianist, and amateur musicians met regularly at his home in Vienna; he had a standing reservation for the whole family at the great Vienna Opera.

Boltzmann was renowned as an excellent lecturer, and some of his classes were attended by the general public as well as by his regular students. He traveled extensively and left a description of a trip to California in 1905 that is full of lively humor. That was his third visit to the United States, a rather unusual number for the times.

Boltzmann wrote many papers, sometimes three or four in a year, that were long and full of laborious calculations. All that scared his readers. For example, Maxwell wrote to his friend Peter Guthrie Tait (1831–1901): "By the study of Boltzmann, I have been unable to understand him. He could not understand me on account of my shortness, and his length is an equal stumbling block for me, thence I am very much inclined to join the glorious company of the supplanters and to put the whole business in about six lines." On the other hand, Boltzmann almost worshipped Maxwell, since the time his Vienna professor Stefan had given him Maxwell's papers on electricity to read and an English grammar to help him out. In telling that story, Boltzmann adds that he also had a dictionary from his father. Boltzmann repeatedly expressed admiration for Maxwell in lyrical terms. He contributed mightily to the diffusion of Maxwell's ideas on electricity by writing one of the first expositions (1891–1893) of Maxwell's electromagnetic theory. The city of Vienna dedicated a monument to Boltzmann inscribed with his famous formula

$$S = k \log W$$

which expresses the magic of statistical mechanics, just as Maxwell's equations express the magic of electricity (see Appendix 6).

When Boltzmann returned to Vienna after his trip to California, one of his pupils, K. Pzibram, drew this caricature of the professor. (Courtesy of D. Flamm.)

Statistics and Probability Enter Physics

As we have seen, the first law of thermodynamics can be reduced, in the case of a suitable molecular model, to the laws of elastic collisions—in other words, to classical mechanics. It was obviously tempting to try something similar for the second law. The ultimate result was the recognition that this is impossible without the addition of statistical considerations. The path to this recognition, however, was long and twisted.

Boltzmann's monument in Vienna, with the famous formula $S = k \log W$. (Courtesy of D. Flamm.)

In 1866 Boltzmann tried a first attempt; he thought he had obtained a valid result, but it was limited to reversible processes for which the second principle, although valid, is less interesting than in the general case. In 1871 Clausius tried the same approach, without having read Boltzmann's previous work. In 1867, however, Maxwell had tried to give a better proof of his 1859 discovery of the molecular velocity distribution in a gas. In a very substantial paper, he analyzed the collisions between molecules and showed that his distribution would be not altered by them. He did not show that that was the only distribution enjoying that property, but he gave arguments for the surmise that any distribution would in time tend to a Maxwellian. He also applied the velocity distribution to improve the calculation of the viscosity coefficient and other transport coefficients. By then he also knew the difficulties with equipartition and had developed the paradoxes connected with a "being whose faculties are so sharpened that he can follow every molecule in its course."

The next great step was made by Boltzmann in 1871, when he started to distinguish between average in time on a single molecule or instant average over many molecules. The two are the same, but a proof of that fact is fraught with difficulties that even now are not completely removed. Boltzmann succeeded in generalizing Maxwell's law to the case in which there are forces (for instance, gravity) acting on the molecules, and he obtained (in modern notation) the famous formula for the "Boltzmann distribution":

$$dW = \frac{\exp(-E/kT) \, d\omega}{\int \exp(-E/kT) \, d\omega}$$

where dW is the probability (in either sense given above) of finding a point representing a system of n degrees of freedom in a phase space volume $d\omega = dq_1, \ldots dq_n \, dp_1 \ldots dp_n$. Here, $q_1 \ldots q_n$ are coordinates of the system, and $p_1 \ldots p_n$ are their conjugate momenta, and $E(q,p)$ is the energy.

In the meantime, Maxwell had arrived at the conclusion that the second law of thermodynamics was of a new kind, unknown in previous physics. It was a statistical law that could be proved only by probability arguments. Hence, it was not absolutely valid, as, for instance, the conservation of energy, but only so probable that the whole duration of history, nay of the earth from its formation, was insufficient to give an appreciable probability for observing its violation in a macroscopic system. In principle, however, a violation is possible, and Maxwell devised a "Demon" (according to Thomson's denomination) that would do that. He first mentioned it in 1867, and wrote about it to his friend Lord Rayleigh:

Glenlair, Dec. 6th, 1870.

Dear Strutt,—

If this world is a purely dynamical system, and if you accurately reverse the motion of every particle of it at the same instant, then all things will happen backwards to the beginning of things, the raindrops will collect themselves from the ground and fly up to the clouds, etc. etc. and men will see their friends passing from the grave to the cradle till we ourselves become the reverse of born, whatever that is. We shall then speak of the impossibility of knowing about the past except by analogies taken from the future and so on. The possibility of executing this experiment is doubtful, but I do not think it requires such a feat to upset the 2nd law of thermodynamics.

For if there is any truth in the dynamical theory of gases, the different molecules in a gas of uniform temperature are moving with very different velocities. Put such a gas into a vessel with two compartments and make a small hole in the A B wall about the right size to let one molecule through. Provide a lid or stopper for this hole and appoint a doorkeeper very intelligent and exceedingly quick, with microscopic eyes, but still an essentially finite being. Whenever he sees a molecule of great velocity coming against the door from A into B he is to let it through, but if the molecule happens to be going slow, he is to keep the door shut. He is also to let slow molecules pass from B to A but not fast ones. (This may be done if necessary by another doorkeeper and a second door.) Of course he must be quick, for the molecules are continually changing both their courses and their velocities.

In this way the temperature of B may be raised and that of A lowered without any expenditure of work, but only by the intelligent action of a mere guiding agent (like a pointsman on a railway with perfectly acting switches who should send the express along one line and the goods along another). I do not see why even intelligence might not be dispensed with and the thing made self-acting.

Moral. The 2nd law of thermodynamics has the same degree of truth as the statement that if you throw a tumblerful of water into the sea, you cannot get the same tumblerful of water out again.

Similar ideas had been exchanged privately between Loschmidt and Boltzmann.

In 1872 Boltzmann wrote a fundamental paper that is still one of the classics of statistical mechanics. In it he considered an arbitrary distribution in space and velocity for monatomic molecules and derived the equation according to which the distribution would change in time under the action of the collisions. He also built an expression dependent on the distribution function, which he called E and later H (hence the later name, H-theorem) and showed that H could only decrease in time or remain stationary. Boltzmann obtained that result by a deep analysis of the possible collisions, a study of their symmetry properties, and laborious calculations. For the Maxwellian distribution, H is stationary; for all others, it decreases. Ultimately, Boltzmann showed that H was the negative of the entropy. He had thus connected thermodynamics with mechanics, but through the roundabout way of the H-theorem.

Die Ordinaten dieser Curve sind fast ausnahmslos sehr klein und diese kleinen Ordinaten sind natürlich nicht mit Vorliebe Maxima. Lediglich die Ordinaten von ganz ungewöhnlicher Grösse sind es meist und zwar um so wahrscheinlicher, je grösser sie sind. Dass eine sehr grosse Ordinate H_0 öfter einem Maximum als dem Durchschnittspunkte der Geraden $y = H_0$ mit einem noch grösseren Buckel entspricht (l. c. p. 797), kommt von der enormen Zunahme der Seltenheit der Buckel mit wachsender Höhe.

The change in quantity H over time, according to Boltzmann. The graph is symbolic because "zincography could not reproduce the details of a real case." From any maximum, the descent resembles an inverted tree; as we descend, more and more paths become available. (From Annalen der Physik und Chemie, 1897. Courtesy of the Library, University of California, Berkeley.)

There were, however, critical points to be discussed in depth, and, indeed, the proof involves the so-called ergodic hypothesis, on which whole books have been written. At present, some of those discussions have been made obsolete or radically changed by quantum mechanics. The physical solution to the difficulties should take into account the identity of the molecules and should be formulated in the framework of quantum mechanics. Thus, Boltzmann's fierce polemics with Ernst Friedrich Zermelo (1871–1953), a distinguished mathematician, at the time assistant to Planck, are mainly of mathematical interest.

In 1877 Boltzmann finally gave a formulation and a proof of his distribution law that is essentially statistical, almost without any dynamics; that is the proof taught most frequently in elementary courses. It contains a postulate that to an extent masks the difficulties. In spite of this, its method has been adequate for all later developments of statistical mechanics, including its extension to quantum statistics, in which molecules are indistinguishable from each other.

In this type of argument, one must consider distribution not in velocity and coordinate, but in momentum and coordinate, the so-called phase space, first introduced in statistical mechanics by Hewett Cottrell Watson (1804–1881) in 1876. The importance of phase space is due to a theorem discovered by Joseph Liouville (1809–1882), which says that the space occupied by a certain number of representative points does not change in time. In other words, representative points move in phase space as in an incompressible fluid. Furthermore, the representative points are confined to a hypersurface of constant energy because energy is conserved. The fundamental hypothesis is that all volume elements of phase space are equally probable.

Possibly Maxwell had stopped reading Boltzmann's papers because he found them too laborious. He seems to have not thought too much of them, for in a private letter to Tait he wrote:

It is rare sport to see those learned Germans contending for the priority of the discovery that the 2nd law of $\theta\Delta cs$ [thermodynamics] is the Hamiltonsche Princip, when all the time they *assume* that the temperature of a body is but another name for the vis viva of one of its molecules, a thing which was suggested by the labours of Gay Lussac, Dulong, etc., but first deduced from dynamical statistical considerations by dp/dt [i.e. Maxwell]. The Hamiltonsche Princip, the while, soars along in a region unvexed by statistical considerations, while the German Icari flap their waxen wings in nephelo-coccygia [cloud cuckooland] amid those cloudy forms which the ignorance and finitude of human science have invested with the incommunicable attributes of the invisible Queen of Heaven.

However, Boltzmann's elaborate approach through the H-theorem gave as a by-product the Boltzmann diffusion equation, which gives the time change of the distribution function. That equation proved to be the real way of solving problems of viscosity, diffusion, and heat conduction, which are accessible to experiment and with which even Maxwell had found difficulties and committed errors. Boltzmann's thought evolved, and by 1877 he better understood the nature of the time change of the function H and its connection with probability. Loschmidt's criticisms had contributed much to that clarification of ideas. However, up to the time of Boltzmann's death, statistical mechanics retained difficulties, such as the apparent exceptions to the equipartition theorem as manifested by the specific heats. The root of the problem was in quantum theory and, thus, beyond the physics of the times.

Real Gases, as Found in Nature

Perfect gases had been the ideal test objects for theoretical studies of thermodynamics and kinetic theory. They have only a small drawback: They do not exist. For many years precision measurements on real gases had revealed their departure from perfection. The precision measurements of Henri Victor Regnault (1810–1878) showed this, as well as the liquefaction of gases obtained by many experimenters. That last phenomenon showed remarkable peculiarities. In general, high pressure and low temperature brought about the condensation of gases. However, Charles Cagniard de la Tour (1777–1859), a versatile man of many parts, a civil servant, friend of Gay-Lussac, and a member of the French establishment, found in 1822 that there was a limiting temperature above which a gas does not condense, no matter how large the pressure to which it is subjected. The concept of critical temperature was further developed by Thomas Andrews (1813–1885), a remarkable Victorian British scientist. He was born in Belfast, Ireland, the son of a linen merchant. He studied chemistry first at Glasgow and then in Paris, also taking a degree in medicine at Edinburgh. He settled in Belfast as a practicing physician and professor of chemistry at Queens College. He also maintained frequent contacts with German and French colleagues such as Jean-Baptiste-André Dumas (1800–1884), Friedrich Wöhler (1800–1882), Jöns Jacob Berzelius (1779–1848), and Justus von Liebig (1803–1873). In England he was a friend of Faraday, Maxwell, and Tait. Andrews' studies on many subjects of what we would presently call physicochemistry, are now unimportant; he is remembered for his clarification, in part with Tait, of the nature of ozone and, above all, for his investigation of the equation of state of carbon dioxide. He drew careful curves of pressure versus specific volume for this substance (see illustration below).

As usual, two lines of approach led to an understanding of real fluids. Experiment gave the substance of the phenomena to be described and explained; molecular theory supplied the theoretical tool by which one hoped to obtain such a description.

On the experimental side, the problem of liquefying gases proved to be a fertile ground for discovering the essentials of the phenomenology. The liquefaction of gases had made great progress in the first half of the nineteenth century, but several gases, such as oxygen, nitrogen, and hydrogen, no matter how much they were compressed, refused to liquefy. They were then named permanent gases, because it was deemed impossible to liquefy them. The discovery of the critical temperature gave a reason for their behavior and pointed to the necessity of precooling them before applying pressure. Liquid carbon dioxide could be obtained at a temperature below $31\,°C$, but if it was heated above that temperature it became

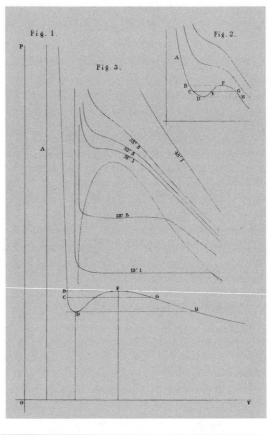

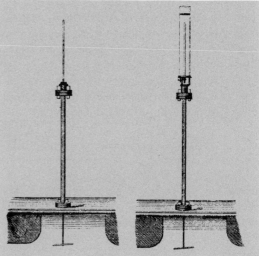

Isothermals of carbon dioxide (top), as measured by Thomas Andrews, together with an illustration of the van der Waals equation and Maxwell's line of stable coexistence of liquid and gas. This famous figure appears in many physics textbooks; the present version is from van der Waals's thesis, as reproduced in Physical Memoirs *(Physical Society of London), 1 (1890), 3. The bottom illustration shows the apparatus used by Andrews for his measurements. (From* Philosophical Transactions, 2 *(1869), 575. Courtesy of the Library, University of California, Berkeley.)*

a gas and the transition was continuous — liquid and gaseous state could not be distinguished above the critical temperature.

By 1877 the art of liquefying gases had progressed to the point where oxygen could be liquefied. In a dramatic series of events, Louis Paul Cailletet (1832–1913), a mining engineer from Châtillon-sur-Seine had announced his success to the Académie des Sciences. He had earlier compressed acetylene, hoping that it would be easier to liquefy than air. The apparatus suddenly sprung a leak, and the gas dropped its pressure adiabatically. A mist formed and Cailletet promptly understood that the adiabatic cooling had condensed the gas. He then successfully applied the same method to oxygen. He performed this work at the beginning of December but communicated it to the Académie only on the twenty-fourth. He had waited because he was a candidate for the Académie and he thought that announcing his discovery just before election time might have been considered in poor taste and might jeopardize his election, which occurred on the seventeenth of December. Unfortunately for him, the secretary of the Académie immediately had to announce that on the twenty-second of December he had received a telegram from Raoul-Pierre Pictet (1846–1929) of Geneva, announcing that he, too, had liquified oxygen on that day. Apparently, Cailletet had been scooped. But no: On the third of the month, Cailletet had written a letter to the famous chemist Henri Étienne Sainte-Claire Deville (1818–1881), telling him of his experiment; Cailletet had also demonstrated it to him. Deville communicated the letter, under the seal of secrecy, to the secretary of the Académie, the same day, thus saving Cailletet's priority.

The competition in liquefaction of gases continued, and significant progress was achieved by Zygmunt Florenty von Wróblewski (1845–1888) and Karol Stanislav Olszewski (1846–1915) in Kraków, Poland. The first had been in Paris as a student and secured a Cailletet apparatus. On his return to Poland, he collaborated with his chemist colleague Olszewski and introduced important improvements by which he could prepare relatively large quantities of liquid oxygen. The collaboration, however, was short lived, and when the Poles, working separately, tried to liquefy hydrogen, they failed. Wróblewski died in a fire in his laboratory in 1888. A new competitor arose in James Dewar (1842–1923), successor to Faraday as director of the Royal Institution. He invented the Dewar flask — the common thermos bottle — which was a truly important step forward in cryogenic techniques. After much effort, Dewar finally liquefied hydrogen in 1898. Now only helium resisted Dewar's efforts. However, Dewar was a difficult man, and he had exchanged severe polemics with Olszewski as well as with English colleagues. He was working practically alone, on a relatively small scale and using artisan methods. The Dutch physicist Heike Kamerlingh-Onnes (1853–1926), on the other hand, introduced sound engineering practice and a truly scientific approach to all low-temperature physics. He was rewarded for his efforts in 1908, when he saw helium boiling in his apparatus. That discovery, too, had its drama, because the helium meniscus was difficult to see, and it took hours before a visitor pointed out to the despondent Kamerlingh-Onnes, who was convinced he had failed, that he had a triumphant success.

Van der Waals's Marvelous Equation

Theory helped greatly the gas experiments by giving a comprehensive description of the states of aggregation of matter. The Dutch physicist Johann Diderik van der Waals combined empirical data, molecular models, thermodynamics, and kinetic theory to formulate an equation of state that was very simple, reasonably accurate, and understandable on a molecular basis.

Van der Waals came from a relatively poor family and could not attend the university until 1862, when he was twenty-five. He earned his living as a secondary school teacher, and ultimately he completed a dissertation for Leyden University in 1873. Dutch dissertations were usually substantial, but van der Waals's was altogether a major work. In it he improved the equation of state for a gas, including a force between molecules and the finite volume of the molecules. He argued that at long distances there must be an attractive force between molecules that acts in addition to the pressure exerted by the walls of the container. He further contributed arguments for the hypothesis that this addition pressure is inversely proportional to the square of the specific volume of the gas. Furthermore, the space available to the molecules must be diminished by the amount occupied by them, or, better, by an amount proportional to the volume occupied by the molecules if they were in contact with each other. The equation of state for one mole of a real gas then becomes

$$\left(p + \frac{a}{V^2}\right)(V - b) = RT$$

That simple equation contains two constants, a and b, that can be determined by experiment for each substance. R is the universal gas constant.

The curves at T constant, the isothermals, are shown for a particular case in the illustration on page 246. They are of two types: At high temperatures they are intercepted by a line $p =$ constant only in one point; at low temperatures they are intercepted in three points. The isothermal that separates the two families has an inflection point with a horizontal tangent. That isothermal is called the critical isothermal, and the inflection point is the critical point. In the limit of high temperature, the isothermals end by coinciding with those of a perfect gas. The low temperature isothermals are, in reality, replaced, for a certain volume interval, by a straight line corresponding to the coexistence of liquid and vapor. In fact, without changing temperature or pressure, a real substance may be all liquid or all vapor or part liquid and part vapor. That situation is represented by the horizontal part of the isothermal. Where should it be located? Maxwell showed by an application of thermodynamics that the criterion to be used is that the two loops determined by the horizontal and the van der Waals isothermal should have the same area.

It is surprising how well van der Waals's equation, with only two empirical constants, reproduces a wealth of data with good approximation.

At the critical point $V_c = 3b$, $p_c = a/27b^2$, and $T_c = 8a/27bR$. It is thus possible to eliminate the constants from the equation by using as variables $p/p_c = \pi$, $v/v_c = \phi$, and $T/T_c = \theta$. The equation of van der Waals then takes the form

$$\left(\pi + \frac{3}{\phi^2}\right)(3\phi - 1) = 8\theta$$

That formulation expresses the law of corresponding states. It was used extensively for exploratory work, especially in the great Leyden laboratory.

It is not easy today to appreciate the importance of van der Waals's work. By now we know so much about molecules that his results appear primitive and perhaps naive, but at the time Maxwell and Boltzmann were deeply impressed by them. Boltzmann devoted a large portion of his book on kinetic theory to the work of van der Waals and called him the "Newton for the deviation of gases from the Boyle law," just as Maxwell had called Ampère the "Newton of electricity." Van der Waals spent the rest of his life improving his thesis, and I say that without irony, because the thesis was so full of new, important ideas.

The Yankee Physicist Gibbs

An appropriate end to the consideration of thermodynamics and statistical mechanics in the nineteenth century requires a brief mention of Josiah Willard Gibbs (1839–1903). That singular figure signals the entry of the United States into theoretical physics. The American Republic had had distinguished physicists: Joseph Henry (1797–1878) and Henry Augustus Rowland (1848–1901) are possibly the most eminent in the nineteenth century besides Gibbs. Their contributions were important but not capital, and their impact on physics was relatively modest. Henry was a contemporary of Faraday and successfully worked on electromagnetic induction, electric motors, and other subjects at the borderline between electrical engineering and physics. He was also an excellent organizer of science. The greatest achievement of Rowland, or at least the one with the widest repercussions, was the building of diffraction gratings of superior quality. They were unequaled for many years and produced true progress in optics and spectroscopy. For many reasons intellects of the United States were more inclined to experimental or even technological endeavors. Gibbs is the exception; his abstract work is "pure theory."

Gibbs was the son of a Yale professor of sacred literature. His roots were sunk deeply in the New England soil. His father and mother died in 1855 and 1861, respectively, and Josiah was left with four sisters. Although in a comfortable financial situation, the unusual family composition took its toll, and two sisters died young. Josiah addressed his studies to engineering and obtained the second Ph.D. awarded in the United States. He started teaching at Yale and took some patents in mechanical devices. In 1866 he went with his two surviving sisters to Europe. He spent time in France and in Heidelberg and Berlin, although he was plagued by ill health. He learned much from the courses he attended, but he did not personally meet Clausius or Maxwell, who would have been of direct interest to him.

On his return to Yale, he was appointed professor of mathematical physics without salary, on the pretext that he did not need it. That strange procedure did not alienate him from Yale. It was only when Johns Hopkins University, where

Josiah Willard Gibbs (1839–1903), professor of mathematical physics at Yale University, was the foremost American theoretical physicist of his time. (Courtesy of AIP Niels Bohr Library.)

Rowland was establishing the best physics department in the United States, offered him a chair, that Yale decided to pay Gibbs. In the meantime, Gibbs composed a vast, original memoir on thermodynamics entitled "On the Equilibrium of Heterogeneous Substances," which he published in the *Transactions of the Connecticut Academy of Sciences*. The long paper contained important original ideas; it was written with the utmost economy of words and in a rather abstract way. Thus, it was inaccessible for two reasons—its unusual place of publication and its intrinsic difficulty. Gibbs sent reprints to the most famous contemporary physicists, and Maxwell and van der Waals read it and appreciated it, but I doubt that there were many others besides them and their students or friends. As a consequence, it was many years before European physicists rediscovered Gibbs's results. That happened to Helmholtz, Planck, and Einstein, among others, to their disappointment.

Gibbs's second capital contribution was to statistical mechanics. In a small book, *Elementary Principles of Statistical Mechanics* (1901), he recast the theory in a very general and elegant form. The result was a book that even Jules Henri Poincaré (1854–1912) found difficult to digest, but which had permanent influence on statistical mechanics and is still studied today.

Today the statistical component of physics has branched and flowered. Pure mathematics has made great strides in the study of classical mechanics in complicated systems. Information theory and computer science have found ample regions of

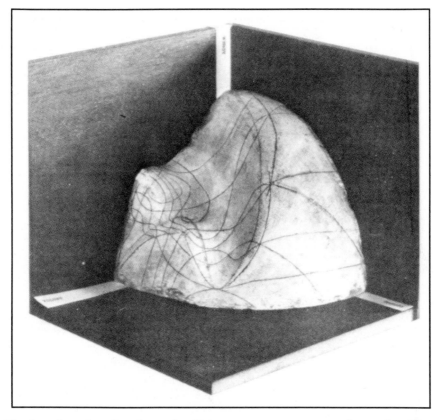

The thermodynamic surface of water in which z = *energy,* y = *entropy, and* x = *volume. From its geometry one can derive all properties of the substance water. The model was executed by Maxwell on the basis of a paper by Gibbs and presented by Maxwell to Gibbs. It has been poetically called "the statue of water." (Courtesy of the Cavendish Laboratory, Cambridge University.)*

contact and influence each other powerfully. Boltzmann would be surprised, and pleased, if he could see these developments. Even biology is deeply affected. It is too early to guess what will emerge from this brew of ideas, some not at all clear, some deep but explored only in a fragmentary way. It is possible that they contain the seeds of key concepts on the working of the mind and on evolution.

Conclusions

I conclude this description of classical physicists and their discoveries with some general remarks.

At the beginning of this book, we find the founding fathers, first of all, Galileo, who developed the modern way of looking at natural phenomena. By now we are accustomed to their methods and teachings, and we do not always appreciate their import and their limitations. If we look at a child getting acquainted with his surroundings, we can conceive of the different possible ways, often animistic and supernatural, in which the mind can interpret sensory data. We can then appreciate the long way mankind traveled to reach the present stage of scientific thought. The fact that the world is rationally intelligible, as Einstein said, is the greatest discovery made by mankind. It is, however, also an act of faith. I have not touched on those questions, nor on the period before Galileo. Although of capital importance, these subjects are beyond the scope of this book.

By the end of the nineteenth century, physics had scored brilliant successes, of which the practitioners were justly proud and confident. They had established methods of experimental investigation and patterns of thought that were adequate to treat an immense number of problems of the macroscopic world. The theories were of several kinds. The paramount examples were mechanics, in its most sophisticated Lagrangean form, able to handle themes as different as celestial mechanics and vibrations of a fluid; field theories, as exemplified by Maxwell's electromagnetism; the independent fief of thermodynamics; and, finally, statistical theories (such as the kinetic theory of gases), which introduced probability into physics, touched the foundations of thermodynamics, and provided a first glimpse of the molecular world.

All this formed essentially a physics of the macroscopic world, of objects of human dimensions; it did not contain an analysis of the world of the atom or of the structure of matter. Innumerable coefficients experimentally determined, such as density, specific heat, resistivity, compressibility, refractive index, electric and magnetic permeability, and so on had to be measured and were then introduced in the equations that describe macroscopic phenomena. The determination of such constants was one of the major occupations of the physicists and required both refined techniques, reminiscent of those of exquisite artisans, and the spirit of the naturalist or (as Rutherford said in a deprecatory sense) of the stamp collector.

One can form an idea of the state of physics around 1890 by consulting some of the great systematic treatises of that time or the table of contents of a major journal, such as *Annalen der Physik* or *Philosophical Magazine*. Perusing one of the major German treatises, we find innumerable detailed drawings of apparatus and descriptions of measuring methods. The use of mathematics is relatively limited. Electromagnetic theory of light is barely mentioned, and I find the name of Regnault mentioned much more frequently than that of Maxwell. Such treatises, of course, do not represent the physics in the minds of a Lorentz, a Poincaré, or a Lord Rayleigh, but they give an impression of what a good university professor knew and what he may have taught.

It also seems that physics had fewer general ideas and guidelines then than now. The skeleton was weaker and the flesh heavier. Of course, this happens in all scientific disciplines: They become leaner. The reason is that since the human learning capacity is more or less constant but the amount of material to be learned increases with time, because of the cumulative nature of science, we reach an impasse. One solution to the quandry is ever-increasing specialization, which, however, has serious drawbacks. Another solution is to find descriptions and teaching methods of greater generality and a more synthetic character. The latter solution should permit students to more rapidly reach the boundary of science, where growth occurs. Indubitably this type of evolution in teaching and learning represents progress, although it entails dropping many subjects that are superfluous or that migrate to other disciplines. The tree of science needs pruning, as real trees do. A consequence of all this is that many of the wonders of the past that were proudly displayed in physics books — the telephone, the telegraph, electric light, photographic objectives — have passed from physics to engineering. In spite of all this, a standard German treatise that around 1890 occupied three or four volumes filled ten times as much space in its 1925 edition.

In the physics treatises, one also finds a trace of the legitimate pride for the technological progress that resulted from scientific discoveries. At the time of Napoleon (i.e., around 1800) the means of transportation, communication, night lighting, prime motors, and other aids to everyday life were not very different from those of the ancient Romans. One hundred years later, mankind had railroads, telegraphs, electric lighting, electric motors, and dynamos. It is no wonder that such achievements should have excited the wave of optimism that pervaded the popularizers of science at the turn of the century.

For the science of physics itself, some of the most far-seeing practitioners, such as Lord Kelvin, mentioned "small clouds on the horizon." One was the failure of the energy equipartition theorem, harbinger of the quantum. Another was the result of the Michelson-Morley experiment on the speed of light, harbinger of relativity. Although Einstein apparently did not know its result until after 1905, ether modelists were puzzled by this experiment. In any case, Hertz and Lorentz had known for some time that electrodynamics of moving bodies presented serious problems, although they did not suspect that their solution would entail such radical measures as proposed by Einstein in his novel analysis of space and time.

All those contradictions concern macroscopic physics, but the times were becoming ripe for a serious study of the atom, beyond the simple, fantastic speculations of the past. There had been numerous attempts to guess at atomic structure. Kinetic theory had given at least the order of magnitude of the mass and dimensions of molecules. Helmholtz had remarked that the laws of electrolysis almost compelled the assumption of an atomic structure of electricity (the electron). It had not escaped Maxwell that the identity of atoms was a peculiar phenomenon that required an explanation. Faraday, with his fertile imagination, had tried to discover how an external magnetic field alters the light emitted by an atom placed in the field, an effect detected by Zeeman in 1896. The study of vacuum discharges, to which many nineteenth-century physicists, such as Faraday, Pluecker, Crookes, Lenard, and others had contributed, evolved with improving vacuum techniques and at the end of the century had given two major results: the isolation of cathode rays (that is, of electron beams) and the discovery of x rays. In a dramatic series of events, the totally unexpected discovery of radioactivity, combined with the discoveries mentioned above, opened the door to the atomic world.

I have told the subsequent story, mainly the exploration of the atom and its nucleus and the still unraveling progress in the subnuclear realm, in *From X-Rays to Quarks*. The events treated there happened to a large extent during my lifetime, and I have met several of the chief actors.

Of the scientists mentioned in this book, Maxwell and Hertz died young, otherwise they would have seen both relativity and the quantum. Boltzmann and Lord Kelvin were able to see the beginnings of both theories, as well as the discovery of radioactivity.

Newton's Mathematical Principles (Section II): The Determination of Centripetal Forces

"Proposition I. Theorem I

The areas which revolving bodies describe by radii drawn to an immovable centre of force do lie in the same immovable planes, and are proportional to the times in which they are described.

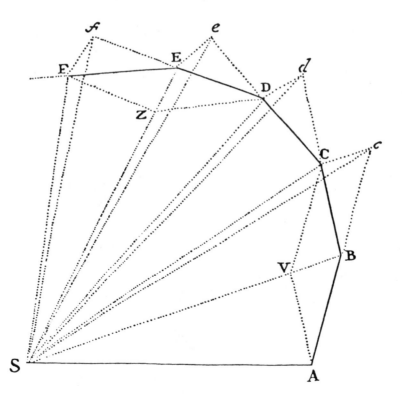

For suppose the time to be divided into equal parts, and in the first part of that time let the body by its innate force describe the right line AB. In the second part of that time, the same would (by Law I), if not hindered, proceed directly to c, along the line Bc equal to AB; so that by the radii AS, BS, cS, drawn to the centre, the equal areas ASB, BSc, would be described. But when the body is arrived at B, suppose that a centripetal force acts at once with a great impulse, and, turning aside the body from the right line Bc, compels it afterwards to continue its motion along the right line BC. Draw cC parallel to BS, meeting BC in C; and at the end of the second part of the time, the body (by Cor. I of the Laws) will be found in C, in the same plane with the triangle ASB. Join SC, and, because SB and Cc are parallel, the triangle SBC will be equal to the triangle SBc, ad therefore also to the triangle SAB. By the like argument, if the centripetal force acts successively in C, D, E, &c., and makes the body, in each single particle of time, to describe the right lines CD, DE, EF, &c., they will all lie in the same plane; and the triangle SCD will be equal to the triangle SBC, and SDE to SCD, and SEF to SDE. And therefore, in equal times, equal areas are described in one immovable plane: and, by composition, any sums SADS, SAFS, of those areas, are to each other as the times in which they are described. Now let the number of those triangles be augmented, and their breadth diminished *in infinitum;* and (by Cor. IV, Lem. III) their ultimate perimeter ADF will be a curved line: and therefore the centripetal force, by which the body is continually drawn back from the tangent of this curve, will act continually; and any described areas SADS, SAFS, which are always proportional to the times of description, will, in this case also, be proportional to those times. Q.E.D.''

For comparison the same proposition in modern notation is proved as follows. The fundamental law of mechanics is

$$\mathbf{F} = m\ddot{\mathbf{r}},$$

where $\mathbf{F}$ is the force applied to the point of mass m and coordinate $\mathbf{r}$. $\mathbf{F}$ and $\mathbf{r}$ are vectors, and the dots denote time derivative. $\mathbf{F}$ by hypothesis has the form $f(r)\mathbf{r}$ where $f(r)$ is an arbitrary function of r.

The angular momentum, by definition is $m\dot{\mathbf{r}} \times \mathbf{r}$ and its time derivative is

$$\frac{d}{dt} m\dot{\mathbf{r}} \times \mathbf{r} = m\ddot{\mathbf{r}} \times \mathbf{r} + m\dot{\mathbf{r}} \times \dot{\mathbf{r}}$$

$$= f(r)\mathbf{r} \times \mathbf{r} + m\dot{\mathbf{r}} \times \dot{\mathbf{r}} = 0,$$

because for any vector $\mathbf{v}$, $\mathbf{v} \times \mathbf{v} = 0$. Hence

$$m\dot{\mathbf{r}} \times \mathbf{r} = \text{constant} \quad \text{Q.E.D.}$$

Newton's Mathematical Principles (Section III): The Motion of Bodies in Eccentric Conic Sections

"Proposition XI. Problem VI

If a body revolves in an ellipse; it is required to find the law of the centripetal force tending to the focus of the ellipse.

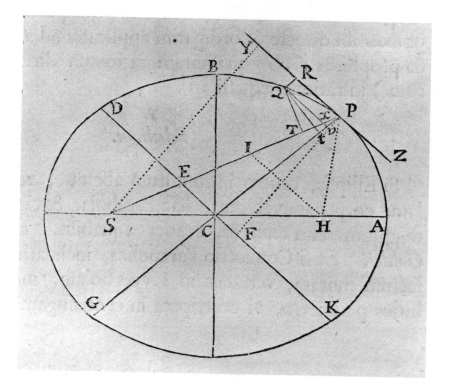

Let S be the focus of the ellipse. Draw SP cutting the diameter DK of the ellipse in E, and the ordinate Qv in x; and complete the parallelogram QxPR. It is evident that EP is equal to the greater semiaxis AC: for drawing HI from the other focus H of the ellipse parallel to EC, because CS, CH are equal, ES, EI will be also equal; so that EP is the half-sum of PS, PI, that is (because of the parallels HI, PR, and the equal angles IPR, HPZ), of PS, PH, which taken together are equal to the whole axis 2AC. Draw QT perpendicular to SP, and putting L for the principal latus rectum of the ellipse $\left(\text{or for } \dfrac{2BC^2}{AC}\right)$, we shall have

$$L \cdot QR : L \cdot Pv = QR : Pv = PE : PC = AC : PC,$$
$$\text{also, } L \cdot Pv : Gv \cdot Pv = L : Gv, \text{ and, } Gv \cdot Pv : Qv^2 = PC^2 : CD^2.$$

By Cor. II, Lem. VII, when the points P and Q coincide, $Qv^2 = Qx^2$, and Qx^2 or $Qv^2 : QT^2 = EP^2 : PF^2 = CA^2 : PF^2$, and (by Lem. XII) $= CD^2 : CB^2$. Multiplying together corresponding terms of the four proportions, and simplifying, we shall have

$$L \cdot QR : QT^2 = AC \cdot L \cdot PC^2 \cdot CD^2 : PC \cdot Gv \cdot CD^2 \cdot CB^2 = 2PC : Gv,$$

since $AC \cdot L = 2BC^2$. But the points Q and P coinciding, 2PC and Gv are equal. And therefore the quantities $L \cdot QR$ and QT^2, proportional to these, will be also equal. Let those equals be multiplied by $\dfrac{SP^2}{QR}$, and $L \cdot SP^2$ will become equal to $\dfrac{SP^2 \cdot QT^2}{QR}$. And therefore (by Cor. I and V, Prop. VI) the centripetal force is inversely as $L \cdot SP^2$, that is, inversely as the square of the distance SP. Q.E.I.

The Same Otherwise

Since the force tending to the centre of the ellipse, by which the body P may revolve in that ellipse, is (by Cor. I, Prox. X) as the distance CP of the body from the centre C of the ellipse, let CE be drawn parallel to the tangent PR of the ellipse; and the force by which the same body P may revolve about any other point S of the ellipse if CE and PS intersect in E, will be as $\dfrac{PE^3}{SP^2}$ (by Cor. III, Prop. VII); that is, if the point S is the focus of the ellipse, and therefore PE be given as SP^2 reciprocally. Q.E.I.

With the same brevity with which we reduced the fifth Problem to the parabola, and hyperbola, we might do the like here; but because of the dignity of the Problem and its use in what follows, I shall confirm the other cases by particular demonstrations.''

Appendix 3

Kepler's Laws in Modern Standard Derivation

Kepler's laws and their Newtonian explanation in modern language can be summarized as follows: Call the mass of the sun M and the mass of a planet m. Call r the distance between the sun and the planet, and θ the angle between r and a fixed direction in the plane of the motion. We skip the simple proof that the motion occurs in a plane. We also assume m/M so small that the sun may be considered to be fixed in space.

The force between the sun and the planet is an attraction of magnitude

$$F = f\frac{mM}{r^2} \tag{1}$$

with $f = 6.67210^{-8}$ din cm^2 g^{-2}.

The motion of the planet around the sun obeys the conservation of energy,

$$\frac{m}{2}(\dot{r}^2 + r^2\dot{\theta}^2) - f\frac{mM}{r} = E \tag{2}$$

and the conservation of angular momentum,

$$mr^2\dot{\theta} = L, \tag{3}$$

where E is the energy and L the angular momentum.

By introducing the abbreviations $c = L/m$, $k = fM$, and $E' = E/m$, and using the identities

$$\dot{r} = \frac{dr}{d\theta}\dot{\theta}, \qquad \dot{\theta} = \frac{c}{r^2} \qquad \dot{r} = \frac{c}{r^2}\frac{dr}{d\theta}, \tag{4}$$

we derive from equations (2) and (3)

$$\left(\frac{d\frac{1}{r}}{d\theta}\right)^2 = \frac{2E'}{c^2} + \frac{2k}{c^2 r} - \frac{1}{r^2}, \tag{5}$$

which is the differential equation of the orbit. Calling $1/r = u$, this equation is transformed into

$$\left(\frac{du}{d\theta}\right)^2 = \frac{2E'}{c^2} + \frac{2k}{c^2}u - u^2. \tag{6}$$

We add and subtract k^2/c^4 to the right side and obtain

$$\left(\frac{du}{d\theta}\right)^2 = \frac{2E'}{c^2} + \frac{k^2}{c^4} - \left(u - \frac{k}{c^2}\right)^2 = q^2 - \left(u - \frac{k}{c^2}\right)^2 \tag{7}$$

with

$$q^2 = \frac{2E'}{c^2} + \frac{k}{c^4}$$

This equation gives

$$\frac{du}{d\theta} = \sqrt{q^2 - \left(u - \frac{k}{c}\right)^2} \quad \text{or} \quad \frac{du}{\sqrt{1 - \frac{1}{q^2}\left(u - \frac{k}{c}\right)^2}} = qd\theta, \tag{8}$$

and by integrating,

$$u = \frac{1}{r} = \frac{k}{c^2} + q\cos(\theta - \theta_0) \tag{9}$$

By a suitable choice of axes the constant of integration θ_0, can be chosen as zero and equation (9) solved for r gives

$$r = \frac{c^2/k}{1 + \frac{c^2 q}{k}\cos\theta}, \tag{10}$$

which is the equation in polar coordinates of a conic of eccentricity $e = c^2 q/k$. We remind that the definition of eccentricity is $e = \sqrt{1 - b^2/a^2}$, where a and b are the major and minor semi axes of the conic. The parameter of the conic is $p = b^2/a = c^2/k$. For $E < 0$, the conic is an ellipse. The polar axis coincides with the major axis of the conic, and the pole is in a focus of the conic. Equation (10) shows that the orbit is an ellipse with the sun at one focus (first Kepler law); equation (3) is in essence the second Kepler law. The third Kepler law is obtained by calculating the period T of the motion. From equation (3) we have $2\pi ab/T = c$. Squaring this expression and dividing by p, we have the third Kepler law in the form

$$\frac{4\pi^2 a^3}{T^2} = \frac{c^2}{p} = fM = \text{constant.} \tag{11}$$

For the earth-sun system $M = 1.991 \times 10^{30}$ kg; $m = 5.979 \times 10^{24}$ kg; $e = 0.00167$; semimajor axis of revolution $= 149.57 \times 10^6$ km; period of revolution $= 3.1558150 \times 10^7$ sec.

Appendix 4

Kirchhoff's Law on Heat Exchange

A surface s at temperature T emits power in the form of electromagnetic radiation at the rate $e(v, T) \, dv$ in the frequency interval dv and per unit surface. The function $e(v, T)$ depends on the nature of the emitting body; for instance, a sodium flame will have a great spike at the frequency corresponding to the yellow D lines; a red-hot iron surface will have a continuum distribution of $e(v, T)$ and at 1,000 K the continuum will peak around a wavelength of about 1 micron. Increasing T will always increase $e(v, T)$ at constant v. The dimensions of $e(v, T)$ are power $\times$ time/surface or m/t^2.

The same surface absorbs at least part of the electromagnetic energy falling on it in the frequency interval dv. Call the fraction of the power absorbed $a(v, T)$. Obviously, $a(v, T)$ is a pure number between 0 and 1.

Consider now two surfaces at the same temperature facing each other and emitting only from the surfaces facing each other. Surface 1 sends power to surface 2 at the rate $e_1(v, T)dv \, S_1$, of which the fraction $a_2(v, T)dv \, S_2$ is absorbed. Similarly, surface 2 sends power $e_2(v, T)dv \, S_2$ to surface 1, of which the fraction $a_1(v, T) \, S_1$ is absorbed. Assume $S_1 = S_2$. The products $e_1(v, T) \, a_2(v, T)$ and $e_2(v, T) \, a_1(v, T)$ must be equal because the two surfaces are at the same temperature, and one cannot get hotter than the other without violating the second principle of thermodynamics in its Kelvin formulation. It follows that

$$\frac{e_1(v, T)}{a_1(v, T)} = \frac{e_2(v, T)}{a_2(v, T)} = u(v, T), \tag{1}$$

where $u(v, T)$ is a universal function of v and T independent of the substances considered.

Kirchhoff's law, that is equation (1), explains such phenomena as the inversion of spectral lines. At the same time it poses the problem of finding $u(v, T)$. Experimentally, $u(v, T)$ can be determined by studying the emission spectrum of a body having $a(v, T) = 1$, a black body. This is practically realized by an enclosure with walls at the temperature T and a small hole from which the radiation escapes.

The investigation of $u(v, T)$ brought the discovery of Planck's constant h and the inauguration of quantum physics, the portal to modern physics.

Appendix 5

The Arguments of the "Newton of Electricity"

Ampère, in trying to emulate Newton, wanted to set up law for the interaction between two elements of current $i_1 dl_1$ and $i_2 dl_2$. The intensities of the currents in the two circuits are, respectively, i_1 and i_2 measured in an arbitrary unit. Ampère was well aware that the interaction between elements of current is unobservable because only closed circuits are accessible to experiment. Nevertheless, he searched for an elementary law that would give the right answer in all observable cases; furthermore, he considered it highly desirable that the force between elements of current should be directed along the line joining them. He saw this as a requirement satisfying automatically the principle of equality of action and reaction. He tried to find this elementary force, taking into account the results of his four fundamental experiments. The first and second experiment show that the force of dl_1 on dl_2 is the same as the vector sum of the forces produced by dl_1 on the components of dl_2 and the relation between dl_1 and dl_2 must be symmetrical. To satisfy these conditions, only forms linear in the components of dl_1 and dl_2 may appear. The only possible ones are obtained from $dl_1 \cdot dl_2$ and from $(dl_1 \cdot \hat{r}_{12})(dl_2 \cdot \hat{r}_{12})$, where $\hat{r}_{12}$ is a unit vector in the directions $r_2 - r_1$. The first experiment shows that the force depends linearly on i_1 and i_2. Hence, its most general expression is

$$d\mathbf{F} = i_1 i_2 \hat{r}_{12}[f_1(\mathbf{r}_{12})dl_1 \cdot dl_2 + f_2(\mathbf{r}_{12})(dl_1 \cdot \hat{r}_{12})(dl_2 \cdot \hat{r}_{12})] \tag{1}$$

where $f_1(r_{12})$ and $f_2(r_{12})$ are arbitrary functions of $r_{12} = |\mathbf{r}_1 - \mathbf{r}_2|$.

The fourth experiment shows that the force is unchanged if all lengths are multiplied by the same factor μ and the currents are kept constant. Then the right hand expression is multiplied by μ^2 because of the factors dl_1 and dl_2 and, to maintain the force constant, f_1 must be equal to A/r_{12}^2 and f_2 to B/r_{12}^2.

That the force is thus inversely proportional to the square of the distance between the elements of current can only warm the heart of a Newtonian. The third experiment negates a force parallel to dl_2 produced by any closed circuit l. This requirement imposes a relation between A and B. For instance, a detailed calculation making the closed circuit coplanar with dl_2 and composed of two arcs of circle centered in dl_2 and two radii converging on dl_2, shows that it must be $A = -(\frac{2}{3})B$. The condition is also sufficient.

In conclusion, equation (1) becomes

$$dF = i_1 i_2 \frac{\hat{r}_{12}}{r_{12}^2} A \left[(dl_1 \cdot dl_2) - \frac{3}{2} (dl_1 \cdot \hat{r}_{12})(dl_2 \cdot \hat{r}_{12}) \right], \qquad (2)$$

with the constant A depending upon the choice of units.

Equation (2) is Ampère's formula, a completely acceptable form of the elementary force.

This form is not unique because, as we have pointed out, any other form that gives the same force between closed circuits would be equally acceptable; this means that we can add any quantity that vanishes upon integration on closed circuits. Taking advantage of this arbitrariness, equation (2) has been transformed in many ways. This was a favorite sport around 1850 and led to some important generalizations.

A particularly symmetric form obtainable is

$$dF = \frac{i_1 i_2}{r_{12}^2} \left[(dl_1 \cdot \hat{r}_{12})dl_2 + (dl_2 \cdot \hat{r}_{12})dl_1 - (dl_1 \cdot dl_2)\hat{r}_{12} \right]. \qquad (3)$$

It will be noted that here the elementary force is not directed along the line joining dl_1 and dl_2, an arbitrary assumption that entered equation (2).

In equation (3) the first term does not contribute to the integral because it can be written as $\oint \text{grad} \frac{1}{r_{12}} dl_1$, which is nil by the properties of the integral of a gradient on a closed path. The other terms under integration can be transformed, using the vector identity $a \times (b \times c) = (a \cdot c)b - (a \cdot c)c$ into the form

$$F = \oint_2 i_2 \oint_1 i_1 \frac{dl_2 \times (dl_1 \times \hat{r}_{12})}{r_{12}^2}. \qquad (4)$$

This, had he grasped its mathematics, would have pleased Faraday. He would have at once read it as one of the currents, say circuit 1, producing a magnetic field B and, rather than perform integrals, he would have seen the lines of force and verified their presence with iron filings. Of course, a magnetic field acts on an element of current according to a force perpendicular to both current and field. The result of equation (4) is thus interpreted by the generation of field B as

$$B = i_1 \oint_{l_1} \frac{dl_1 \times \hat{r}_{12}}{r_{12}^2}; \qquad (5a)$$

and the action of B on the element of current as

$$dF = i_2 dl_2 \times B. \qquad (5b)$$

We are educated in the tradition of Faraday, which is much more powerful and simpler than Ampère's approach. This Appendix should show the relations between the two points of view.

Appendix 6

The Measurement of the Ratio of Electrostatic to Electromagnetic Units of Charge and the Velocity of Light

At page 165 we reported Maxwell's words on the definition of electrostatic and electromagnetic units. Here we see an implementation of his program. We use cgs units of length, mass, and time.

The fundamental law by which we define the electrostatic units is that two equal point charges q at distance r repel each other with a force $F = q^2/r^2$, or that one charge generates a field $\mathbf{E} = q/r^2$. These equations give the dimensions of $q = l^{3/2}m^{1/2}t^{-1}$ and of $\mathbf{E} = l^{-1/2}m^{1/2}t^{-1}$.

The fundamental law by which we define the electromagnetic units is Ampère's law of force, equation (2) in Appendix 5, with $A = 1$. It gives for the force per unit length of two rectilinear parallel circuits at distance r, $F/l = 2i^2/r$, or, for the field at the center of a circular loop of radius r, $\mathbf{B} = 2\pi i/r$. The dimensions of i are thus $l^{1/2}m^{1/2}/t$ and, hence, those of q are $l^{1/2}m^{1/2}$. The dimensions of $\mathbf{B}$ in electromagnetic units are the same as those of $\mathbf{E}$ in electrostatic units. The ratio q_{es}/q_{em} has the dimensions of a velocity.

To determine this ratio, W. Weber and R. Kohlrausch in 1856 measured a charge in the two systems. They took a condenser and compared its capacity with that of a sphere of radius R. The capacity of the sphere is R; that of the condenser is C, both in cm. The voltage on the condenser can be measured on an absolute electrometer. This is essentially a balance with which one measures the force F between two plates of surface S at distance d. One has $V = \sqrt{8\pi \dfrac{Fd^2}{S}}$. The charge on the condenser is $Q = CV$, and by performing the operations described, it is measured in electrostatic units.

The condenser is now discharged through a loop of radius ρ situated in the plane of the magnetic meridian. At the center of the loop there is a magnetic needle of magnetic moment μ and moment of inertia I. When no current is sent through the loop, the needle oscillates around it, its equilibrium position. The angle of oscillation θ, assumed to be small, obeys the homogeneous differential equation

$$\ddot{\theta} + \frac{B_0\mu}{I}\theta = 0 \tag{1}$$

and the frequency of the oscillation is $v = \frac{1}{2\pi}\sqrt{\frac{B_0\mu}{I}}$, where B_0 is the horizontal component of the earth's magnetic field. We measure v, and, as we will need the value of B_0, we shall determine it by a separate experiment.

When the condenser is discharged through the loop with the needle initially at rest, we can assume that the discharge occurs in a time τ extremely short compared with the period of oscillation and that during that time the needle does not move appreciably from the equilibrium position (ballistic galvanometer). After the discharge, still at time practically zero, the needle has acquired an angular momentum $B\mu\tau = \frac{2\pi}{\rho}i\mu\tau = \frac{2\pi}{\rho}Q\mu = I\dot{\theta}(0)$, where B is the average value of the magnetic field during τ. The integral of equation (1) with $\theta(0) = 0$ and $\dot{\theta}(0) = \frac{2\pi Q\mu}{\rho I}$ is

$$\theta = \frac{Q\mu}{\rho Iv}\sin 2\pi vt. \tag{2}$$

We measure the amplitude of the oscillation A, and we obtain, using the value of the frequency,

$$A = \frac{Q\mu}{\rho Iv} = \frac{4\pi^2 vQ}{\rho B_0} \quad \text{or} \quad Q = \frac{B_0\rho}{4\pi^2 v}A. \tag{3}$$

The auxiliary measurement of B_0 can be made by a method that goes back to Gauss. A magnetic bar of moment of inertia G, measured from its weight and dimensions, is made to swing around its equilibrium position in the north-south direction. The frequency of oscillation is

$$v_0 = \frac{1}{2\pi}\sqrt{\frac{B_0 M}{G}}, \tag{4}$$

where M is the unknown magnetic moment of the bar. The same magnet produces a field of intensity $B' = 2M/r^3$ at a point located at a distance r from its center, on a line perpendicular to the direction of the bar. This field is directed parallel to the bar. If the magnetic bar is now placed in the east-west direction, this field is perpendicular to B_0, and the field resulting from it and B_0 makes an angle α with the magnetic meridian given by

$$\tan \alpha = 2M/r^3 B_0. \tag{5}$$

By measuring this angle and using equations (4) and (5), we obtain the value of

$$B_0 = \left(\frac{8\pi^2 v_0^2 G}{r^3 \tan \alpha}\right)^{1/2}.$$

Now everything has been reduced to measurements of lengths, masses, and times, and we can have Q in electromagnetic units.

The implementation of the measurements requires many refinements that we omit. However, already in 1856 the result of Weber and Kohlrausch was $q_{em}/q_{es} = 3.107 \times 10^{10}$ cm/sec, comparable to the measurement of c by Fizeau giving 3.14×10^{10} cm/sec.

Appendix 7

Plane Waves from Maxwell's Equations

Maxwell's equations for an insulating isotropic medium of dielectric constant ϵ and permeability μ, in absence of charges and currents, are

$$\text{div } \mathbf{E} = 0 \qquad (1) \qquad\qquad \text{div } \mathbf{H} = 0 \qquad (2)$$

$$\text{curl } \mathbf{E} = -\frac{\mu}{c}\frac{\partial \mathbf{H}}{\partial t} \qquad (3) \qquad\qquad \text{curl } \mathbf{H} = \frac{\epsilon}{c}\frac{\partial \mathbf{E}}{\partial t}, \qquad (4)$$

where $\mathbf{E}$ and $\mathbf{H}$ are measured respectively on absolute electrostatic and electromagnetic units. The factor c has the dimensions of velocity and can be determined as described in Appendix 6. The factor c is required because of the choice of units. In the same units ϵ and μ are pure members, and their value in vacuum is 1.

The simplest solution of Maxwell's equations showing waves is obtained when $\mathbf{E}$ and $\mathbf{H}$ depend only on x and t, but not on y and z. In this case one has for the components of $\mathbf{E}$ and $\mathbf{H}$ from equations (1) to (4)

$$\frac{\partial E_x}{\partial x} = \frac{\partial H_x}{\partial x} = \frac{\partial H_x}{\partial t} = \frac{\partial E_x}{\partial t} = 0 \qquad (5)$$

$$-\frac{\partial E_z}{\partial x} = -\frac{\mu}{c}\frac{\partial H_y}{\partial t} \qquad (6) \qquad\qquad -\frac{\partial H_z}{\partial x} = \frac{\epsilon}{c}\frac{\partial E_y}{\partial t} \qquad (7)$$

$$\frac{\partial E_y}{\partial x} = -\frac{\mu}{c}\frac{\partial H_z}{\partial t} \qquad (8) \qquad\qquad \frac{\partial H_y}{\partial x} = \frac{\epsilon}{c}\frac{\partial E_z}{\partial t}. \qquad (9)$$

From equation (5) it follows that E_x and H_x are constants, and we may assume them to be zero without loss of generality. By taking the derivative with respect to x of equation (6) and the derivative with respect to t of equation (9) and eliminating $\partial^2 H_y/\partial t \partial x$, we obtain

$$\frac{\partial^2 E_z}{\partial x^2} = \frac{\epsilon\mu}{c^2}\frac{\partial^2 E_z}{\partial t^2}. \qquad (10)$$

By the same procedure, we obtain equations for E_y, E_z, and H_z.

Equation (10) has a particular solution

$$E_z = \cos\left(vt \pm \frac{x}{\lambda}\right), \tag{11}$$

provided that the constants v, λ obey the relation

$$\lambda v = c/\epsilon\mu.$$

This can be verified by direct calculation. Equation (11) is the equation of a plane wave propagating in the x direction with the velocity λv because the cosine takes the same value at time and place t_0, x_0 and at time and place x, t provided $vt_0 \pm \dfrac{x_0}{\lambda} =$

$vt \pm \dfrac{x}{\lambda}$ or $x - x_0 = \pm \lambda v(t - t_0)$ which means that the wave travels with the velocity

$$v = \lambda v = c/\epsilon\mu.$$

If we assume that the wave is polarized in the z direction, that is, that its electric vector points in the z direction, we find from equation (8) that H_z is constant and can be put equal to zero. On the other hand, H_z can be directly obtained from equations (6) and (9), and it has the value

$$H_y = \sqrt{\frac{\epsilon}{\mu}}\cos\left(vt \pm \frac{x}{\lambda}\right) = \sqrt{\frac{\epsilon}{\mu}}\,E_z.$$

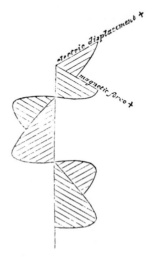

Maxwell's drawing gives "The values of the magnetic force and of the electromotive intensity at a given instant at different points of the ray for the case of a simple harmonic disturbance in one place." (From Treatise on Electricity and Magnetism, *1873.)*

This shows that in a plane wave the electric and magnetic vectors are perpendicular to each other and perpendicular to the direction of propagation of the wave.

In a vacuum the waves have the same amplitude and propagate with velocity c, thus revealing the connection between the ratio of the electromagnetic to the electrostatic unit for the quantity of electricity and the velocity of light. In a medium of dielectric constant ϵ and permeability μ, the refractive index n defined by $c/v = n$ has the value

$$n = \sqrt{\epsilon\mu}.$$

Appendix 8

The Influence of Pressure on
the Melting Point of Ice

One of the earliest applications of thermodynamics, which goes back to Clapeyron and James Thomson, concerns the influence of pressure on the melting point of ice.

One can argue, going back to the original ideas of Carnot, as follows: Make a Carnot engine in which 1 gram of a solid at the melting point melts at temperature T and under pressure p. To melt the solid, we must supply the melting heat r; the solid on melting changes volume, and the volume decreases from v_s to v_l, where v_s and v_l are the specific volumes of solid and liquid. The melting occurs isothermically. We now decrease the pressure adiabatically to $p - dp$, and correspondingly the temperature varies from T to $T - dT$. We then freeze the liquid again isothermically at this temperature, and close the cycle by returning adiabatically to the original conditions.

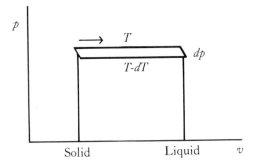

The heat supplied in the first melting, multiplied by the efficiency of the cycle (which according to Carnot is dT/T) gives the work obtained. This work, according to the figure, is $dp\,(v_s - v_l)$. We thus have

$$\frac{dp}{dT} = \frac{1}{T}\frac{r}{v_l - v_s}.$$

This formula is called the Clapeyron formula. Note that the sign of dp/dT is the same as that of $v_l - v_s$ for water, since ice floats on water, $v_l - v_s$ is negative.

In the case of the melting of water, $r = 79.71$ cal/g, which can be transformed to joules at the rate 4.184 joules $= 1$ cal. The specific volume of water is 1.00016; that of ice is 1.0905 in cm³/g; the melting temperature at atmospheric pressure is 273.16 K. Inserting these numbers in the equation, one obtains $dp/dT = -1.351 \times 10^8$ dynes/cm, or -133.36 atmospheres/degree. The melting point decreases at high pressure. In a glacier, the ice at the bottom melts under the pressure of the incumbent mass, and the glacier "flows."

As an example, we will derive Clapeyron's formula also by the standard procedure of expressing that dS is a perfect differential. First, what is a perfect differential? If we have an expression of the type

$$dz = A(x, y)\, dx + B(x, y)\, dy,$$

we say that it is a perfect differential if there is a function $z(x, y)$ such that dz is the differential of such function. A necessary and sufficient condition for this to happen is that

$$\frac{\partial A(x, y)}{\partial y} = \frac{\partial B(x, y)}{\partial x}$$

When this condition is satisfied, it is also possible to integrate dz in the plane x, y, and the integral on a closed path is zero, from which it follows that the integral from a point P to a point Q does not depend on the integration path. The converse of the last property is also true.

All this may be immediately applied to thermodynamics. We may restate the second principle by asserting that there is a function S of the state such that $dS = dQ/T$ is a perfect differential. Note that dQ is NOT a perfect differential.

In our specific case, we choose T and V as variables and write the first and second principles combined as

$$dS = \frac{dU + p\,dV}{T} = \frac{1}{T}\left(\frac{\partial U}{\partial T}\right)_V dT + \frac{1}{T}\left[\left(\frac{\partial U}{\partial V}\right)_T + p\right] dV. \qquad (1)$$

By the property of the total differentials

$$\frac{\partial^2 S}{\partial V \partial T} = \frac{\partial}{\partial V} \frac{1}{T}\left(\frac{\partial U}{\partial T}\right)_V = \frac{\partial}{\partial T}\left[\frac{1}{T}\left(\frac{\partial U}{\partial V}\right)_T + p\right] \qquad (2)$$

and, executing the derivatives and simplifying, we obtain the general equation

$$\frac{1}{T}\left(\frac{\partial U}{\partial V}\right)_T - \left(\frac{\partial p}{\partial T}\right)_V + \frac{p}{T} = 0. \qquad (3)$$

When a unit mass of ice melts at constant temperature, the heat supplied is

$$r = u_w - u_i + p(v_w - v_i).$$

The greatest part goes to the change of internal energy $u_w - u_i$; the remainder goes to external work $p(v_w - v_i)$, where $v_w - v_i$ is the change of volume accompanying the melting.

We thus have

$$\left(\frac{\partial U}{\partial V}\right)_T = \frac{u_w - u_i}{v_w - v_i} = \frac{r - p(v_w - v_i)}{v_w - v_i}$$

which, introduced in the general equation (3), gives Clapeyron's result

$$\frac{r}{v_w - v_i} = T\frac{dp}{dT}.$$

The Absolute Scale of Temperature and the Gas Thermometer

The first principle of thermodynamics says that for a gas

$$dQ = dU + pdV. \tag{1}$$

Gay-Lussac's experiment (expansion of a gas in a vacuum) contemplates a situation in which the external work vanishes because there is no pressure opposing the expansion when the gas expands in a vacuum. The experiment occurs in an isolated system to which no heat is supplied, and hence $dQ = 0$. The experiment shows that the temperature is not changed by the expansion.

Using V and T as variables, equation (1) gives

$$dQ = \left(\frac{\partial U}{\partial V}\right)_T dV + \left(\frac{\partial U}{\partial T}\right)_V dT + pdV = 0, \tag{2}$$

and since dT is by experiment zero (as are dQ and pdV) we must have

$$\left(\frac{\partial U}{\partial V}\right)_T = 0. \tag{3}$$

This relation and the equation of state $pV = RT$ define a perfect gas.

Using the general equation (3) of Appendix 8, which embodies the second principle of thermodynamics, we have that, at constant volume,

$$dp/p = dT/T. \tag{4}$$

On the other hand, in a conventional gas thermometer, the temperature θ is defined by the equation of state, with α constant:

$$pV = p_0 V_0 (1 + \alpha\theta) \tag{5}$$

$$\left(\frac{\partial U}{\partial V}\right)_\theta = 0. \tag{6}$$

At constant volume we have, from equation (5),

$$\frac{dp}{p} = \frac{d\theta}{1/\alpha + \theta},\tag{7}$$

which, compared with the previous expression of dp/p from equation (4) yields

$$T = 1/\alpha + \theta$$

Thus, the gas thermometer gives absolute temperatures.

Appendix 10

Maxwell's Distribution of Velocities of Molecules in His Own Words

Maxwell read a paper at the meeting of the British Association of Aberdeen in September 1859. It was published in the *Philosophical Magazine* in 1860, and it contains the following heuristic argument on the velocity distribution.

"If a great many equal spherical particles were in motion in a perfectly elastic vessel, collisions would take place among the particles, and their velocities would be altered at every collision; so that after a certain time the *vis viva* will be divided among the particles according to some regular law, the average number of particles whose velocity lies between certain limits being ascertainable though the velocity of each particle changes at every collision.

Prop. IV. To find the average number of particles whose velocities lie between given limits, after a great number of collisions among a great number of equal particles.

Let N be the whole number of particles. Let x, y, z be the components of the velocity of each particle in three rectangular directions, and let the number of particles for which x lies between x and $x + dx$, be $Nf(x)dx$, where $f(x)$ is a function of x to be determined.

The number of particles for which y lies between y and $y + dy$ will be $Nf(y)dy$; and the number for which z lies between z and $z + dz$ will be $Nf(z)dz$ where f always stands for the same function.

Now the existence of the velocity x does not in any way affect that of the velocities y or z, since these are all at right angles to each other and independent, so that the number of particles whose velocity lies between x and $x + dx$, and also between y and $y + dy$, and also between z and $z + dz$, is

$$Nf(x)f(y)f(z)dxdydz.$$

If we suppose the N particles to start from the origin at the same instant, then this will be the number in the element of volume $(dxdydz)$ after unit of time, and the number referred to unit of volume will be

$$Nf(x)f(y)f(z).$$

But the directions of the coordinates are perfectly arbitrary, and therefore this number must depend on the distance from the origin alone, that is

$$f(x)f(y)f(z) = \phi(x^2 + y^2 + z^2).$$

Solving this functional equation, we find

$$f(x) = Ce^{Ax^2}, \qquad \phi(r^2) = C^3 e^{Ar^2}.$$

If we make A positive, the number of particles will increase with the velocity, and we should find the whole number of particles infinite. We therefore make A negative and equal to $-\dfrac{1}{\alpha^2}$, so that the number between x and $x + dx$ is

$$NCe^{-x^2/\alpha^2}dx.$$

Integrating from $x = -\infty$ to $x = +\infty$, we find the whole number of particles,

$$NC\sqrt{\pi}\alpha = N, \therefore C = \frac{1}{\alpha\sqrt{\pi}},$$

$f(x)$ is therefore

$$\frac{1}{\alpha\sqrt{\pi}} e^{-x^2/\alpha^2}$$

Whence we may draw the following conclusions:—

1st. The number of particles whose velocity, resolved in a certain direction, lies between x and $x + dx$ is

$$N\frac{1}{\alpha\sqrt{\pi}} e^{-x^2/\alpha^2}dx \qquad (1)$$

2nd. The number whose actual velocity lies between v and $v + dv$ is

$$N\frac{4}{\alpha^3\sqrt{\pi}} v^2 e^{-v^2/\alpha^2}dv \qquad (2)$$

3rd. To find the mean value of v, add the velocities of all the particles together and divide by the number of particles; the result is

$$\text{mean velocity} = \frac{2\alpha}{\sqrt{\pi}} \qquad (3)$$

4th. To find the mean value of v^2, add all the values together and divide by N,

$$\text{mean value of } v^2 = \tfrac{3}{2}\alpha^2. \qquad (4)$$

This is greater than the square of the mean velocity, as it ought to be."

Appendix 11

Boltzmann's Epitaph

Forgetting all the reservations and difficulties besetting the definition of W, the probability of a state, we can show an interesting feature of the relation between probability and entropy, assuming that there is a functional relation,

$$S = f(W). \tag{1}$$

Consider two separate independent systems having entropies S_1 and S_2 and probabilities W_1 and W_2.

Combining the two systems, the entropies add. The composite system has the entropy

$$S = S_1 + S_2. \tag{2}$$

On the other hand, the probability is the product of the probabilities W_1, W_2.

The function S of equation (1) thus has the property that

$$f(W_1) + f(W_2) = f(W_1 W_2), \tag{3}$$

which is satisfied by

$$f(W) = k \log W$$

with k constant.

We thus have Boltzmann's epitaph,

$$S = k \log W.$$

Appendix 12

The Essentials of Boltzmann's
H-theorem

Let $f(\mathbf{r}, \mathbf{v}, t)d\mathbf{r}d\mathbf{v}$ be the probability that at time t a monatomic molecule is contained in a space volume-element and a velocity volume-element around $\mathbf{r}$ and $\mathbf{v}$ in a six-dimensional representative space. The distribution function changes in time for two reasons. First, the motion of the particles, apart from collisions, produces a flow in six-dimensional space $\left(\dfrac{\partial f}{\partial t}\right)_{\text{flow}}$. From analytical mechanics, using Liouville's theorem, one obtains for this, in the absence of forces, $-\mathbf{v} \cdot \text{grad}_r f$, to which one adds a term due to possible potential fields $-\mathbf{a} \cdot \text{grad}_v f$. The other cause of change is collisions, and one has the term $\left(\dfrac{\partial f}{\partial t}\right)_{\text{coll}}$. Collisions can throw out of the volume element certain molecules as well as bringing in others, taking them from all the six-dimensional space.

In the first case $f(\mathbf{v}, \mathbf{r}, t)$ decreases because a molecule of velocity $\mathbf{v}$ collides with one of velocity $\mathbf{v}_1$, and after the collision the two molecules have velocities $\mathbf{v}'$ and $\mathbf{v}'_1$. In the second case, in which $f(\mathbf{v}, \mathbf{r}, t)$ increases, the molecules colliding have before the collision velocities $\mathbf{v}'$ and $\mathbf{v}'_1$, and after the collision velocities $\mathbf{v}$ and $\mathbf{v}_1$. The frequency of collisions of the first kind is proportional to $f f_1$, and the frequency of collisions of the second kind is proportional to $f'_1 f'$. Here and in the integral of equation (1), f'_1 stands for $f(\mathbf{v}'_1, \mathbf{r}, t)$, etc.

The collisional term was calculated by Boltzmann and is the integral in equation (1), which is the famous Boltzmann transport equation

$$\frac{\partial f}{\partial t} + \mathbf{v} \cdot \text{grad}_r f + \mathbf{a} \cdot \text{grad}_v f = \int d\mathbf{v}_1 \int d\Omega g I(g, \theta)[f' f'_1 - f f_1]. \quad (1)$$

Here $\mathbf{a}$ is the acceleration due to an outside potential $U(\mathbf{r})$ that may be present. Clearly, for a molecule of mass m, $\mathbf{a} = -\text{grad}_r U/m$. The indexes on grad denote on which variables one has to take the derivatives.

In the integral, g is the magnitude of the relative velocity of the colliding molecules, before the collision; $I(g, \theta)$ is the appropriate differential collision cross section for collisions producing a deflection θ and for which the final velocity is contained in the solid angle element $d\Omega$.

Boltzmann proved that, for a general $U(\mathbf{r})$, any initial distribution $f(\mathbf{r}, \mathbf{v}, 0)$ will in time change to the stationary Maxwell-Boltzmann distribution, which is time-independent

$$f(\mathbf{r}, \mathbf{v}) = Ae^{-\beta[mv^2/2 + U(\mathbf{r})]}, \tag{2}$$

where A is determined by the total number of molecules that normalize the function and $\beta = 1/kT$. The last equation determining β is obtained by considering the total energy.

Boltzmann's proof proceeds by showing that there is a positive quantity H, which under the influence of the collisions decreases or is stationary

$$\frac{dH}{dt} \le 0. \tag{3}$$

This famous quantity H, for which the theorem is named, is

$$H(t) = \int d\mathbf{r} \int d\mathbf{v} f(\mathbf{r}, \mathbf{v}, t) \log f(\mathbf{r}, \mathbf{v}, t). \tag{4}$$

Ultimately, H will be identified with the entropy, with its sign changed, except for an arbitrary constant; thus, the decrease of H is the increase of entropy brought about by the collisions.

Boltzmann proved that H is constant only if for all collisions

$$f f_1 = f' f_1'. \tag{5}$$

The stationary state by virtue of equation (1) also satisfies the equation

$$\frac{\partial f}{\partial t} + \mathbf{v} \cdot \text{grad}_r f + \mathbf{a} \cdot \text{grad}_v f = 0$$

Equations (1) and (3) determine the Maxwell-Boltzmann distribution for normal cases.

Boltzmann's equation (1) has been studied incessantly since it was formulated in 1872, and it even has practical applications, for instance, to isotope separation. On its centennial, a conference in Vienna was devoted to a review of its status and its consequences. The proceedings of this conference are an eloquent testimonial to the vitality of Boltzmann's equation.

Dilemmas Posed by the Equipartition of Energy

Here is a sample of the paradoxes connected with the equipartition of energy. Consider a perfect gas, either monatomic or polyatomic. Thermodynamics give for the molar heat at constant volume $C_V = \left(\dfrac{dQ}{dT}\right)_V = \left(\dfrac{dU}{dT}\right)_V$ and for the molar heat at constant pressure $C_p = \left(\dfrac{dQ}{dT}\right)_p = \left(\dfrac{dU}{dT}\right)_V + p\left(\dfrac{dV}{dT}\right)_p$. The last term, because of the equation of state, is R. Hence

$$C_p - C_V = R. \tag{1}$$

If molecules have n degrees of freedom, the equipartition theorem requires

$$C_V = nR/2$$

and because of equation (1)

$$C_p/C_V = (n + 2)/n$$

The ratio C_p/C_V can be accurately measured, for instance, from the velocity of sound.

For a monatomic gas, assume 3 degrees of freedom per atom, corresponding to the motion of a material point; one has $C_p/C_V = \frac{5}{2} = 1.667$. Experiments on noble gases support this result. The question may be raised, however, about the degrees of freedom corresponding to the electronic motions in the atom. This difficulty could not be answered by classical physics. Furthermore we may assume that a diatomic molecule is similar to a rigid body shaped like a dumbbell. It then has 6 degrees of freedom, and C_p/C_V should have the value $\frac{8}{6} = 1.3333$. Experiment shows that $C_p/C_V = \frac{7}{5} = 1.40$. For a dumbbell model, we can not deny the 3 degrees of freedom corresponding to the coordinates of the center of mass, nor the 2 coordinates that orient the dumbbell in space. More questionable may appear the sixth degree of freedom, corresponding to the rotation along the line connecting the two atoms forming the dumbbell molecule. Classically, it should certainly be counted, but in fact it is hidden, frozen, as the internal degrees of freedom in the atoms. The reason is to be found in quantum theory.

Appendix 14

The Marvelous Equation of van der Waals and Clausius' Virial Theorem

The equation of van der Waals for N molecules of a real substance is

$$\left(p + \frac{aN^2}{v^2}\right)(v - bN) = NkT \tag{1}$$

and may be inferred qualitatively starting from the perfect gas equation

$$pv = NkT \tag{2}$$

by remarking that the volume accessible to the molecules is not v, but is diminished by the volume occupied by the molecules themselves when they are in contact, Nb. Furthermore, to the pressure exerted by the walls of the gas container we must add a pressure due to attractive forces that may exist between the molecules. This additional pressure is proportional to N^2/v^2 because we may think of it as due to the force acting on a thin surface layer of the gas exerted by the molecules of the bulk of the gas and not counterbalanced on the side of the wall. The number of molecules in the surface layer is proportional to N/v, and the number of attracting molecules is also proportional to N/v; hence the additional pressure is aN^2/v^2.

We may improve this crude qualitative argument using a theorem of Clausius (1870), called the virial theorem.

Given N material points of mass m, we call the virial of the system the quantity

$$V = -\frac{1}{2} \sum m_i \mathbf{r}_i \cdot \mathbf{F}_i, \tag{3}$$

where $\mathbf{F}_i$ is the force acting on the point i of coordinates $\mathbf{r}_i$. Using the equation $\mathbf{F}_i = m_i \ddot{\mathbf{r}}_i$ and the identity

$$\ddot{\mathbf{r}} \cdot \mathbf{r} = \frac{d}{dt} \dot{\mathbf{r}} \cdot \mathbf{r} - \dot{\mathbf{r}}^2,$$

we obtain

$$V = -\frac{1}{2}\frac{d}{dt}\sum m_i\dot{\mathbf{r}}_i \cdot \mathbf{r}_i + \frac{1}{2}\sum m_i\dot{\mathbf{r}}^2.$$

By taking the time average of the virial over a sufficiently long time τ we have

$$\langle V\rangle = -\frac{1}{2\tau}\sum m_i\dot{\mathbf{r}}_i \cdot \mathbf{r}_i\bigg|_0^\tau + \frac{1}{2}\left\langle\sum m_i\dot{\mathbf{r}}_i^2\right\rangle$$

The first right hand term vanishes because the difference between the initial and final values of $\Sigma m_i\mathbf{r}_i \cdot \dot{\mathbf{r}}_i$ stays finite, while τ increases indefinitely. We obtain thus the "virial theorem":

$$\langle V\rangle = \langle E_{\text{kin}}\rangle = \tfrac{3}{2}NkT \tag{4}$$

the last equality is due to the equipartition theorem.

We apply the virial theorem to a gas confined to a cube of side L contained in the first octant of a coordinate system, with sides on the axes and a corner in the origin of the coordinates. In the computation of the virial, it is convenient to distinguish the part due to exterior forces V_e and the part due to intermolecular forces V_i. The virial is the sum of the two. We calculate first V_e. The exterior forces give the pressure on the wall. For the walls perpendicular to the x axis, we have, at abscissa L, $\Sigma F_i L = -pL^3$, while the wall containing the origin does not contribute to the virial because $x = 0$. Extending this argument to all walls and summing the results, we have

$$\langle V_e\rangle = \tfrac{3}{2}pL^3 = \tfrac{3}{2}pv \tag{5}$$

where v is the gas volume.

From equations (4) and (5) we thus have

$$\tfrac{3}{2}NkT = \tfrac{3}{2}pv + \langle V_i\rangle. \tag{6}$$

If there are no intermolecular forces, $\langle V_i\rangle = 0$, and we obtain the perfect gas equation.

We give now the outline of an evaluation of $\langle V_i\rangle$ based on a plausible model, leading to van der Waals's equation. We assume that the force between molecules contains a short-range strongly repulsive zone due to a hard core and a long-range attractive zone where the force varies an r^{-n} with n, in practice, near 6.

Consequently, it is expedient to divide V_i into V_{ic} due to the hard core and V_{ia} due to the attractive force.

To evaluate the part of the virial due to the hard core, we consider the change of momentum due to the collisions. The corresponding short-range and short-duration forces contribute to the virial when the centers of the molecules are at a distance $2r_0$ and are applied with the frequency at which the collisions occur. A calculation gives as final result

$$\langle V_{ic} \rangle = -4(\tfrac{4}{3}\pi r_0^3 N)p = -\tfrac{2}{3}pbN \tag{7}$$

where b, sometimes called the covolume, is 4 times the volume occupied by the molecules (this is the van der Waals constant b).

The contribution to the virial made by the long-range attractive forces $f(r)$ is for each pair of molecules with their centers at distance $> 2r_0$, $-f(r) \cdot r$. Multiplying this quantity by the number of such pairs $4\pi r^2 dr \cdot (N/v) \cdot \tfrac{1}{2}N$ and integrating,

$$v_{ia} = 2\pi \frac{N^2}{v} \int_{2r_0}^{\infty} f(r)r^3 dr \tag{8}$$

Using now equations (6), (7), and (8), we have

$$NkT = pv - pbN + \frac{N^2}{v} \frac{2\pi}{3} \int_{2r_0}^{\infty} f(r)r^3 dr \tag{9}$$

Calling

$$a = \frac{2\pi}{3} \int_{2r_0}^{\infty} f(r)r^3 dr$$

and neglecting small terms containing ab, equation (9) can be rewritten in the van der Waals form:

$$NkT = (v - bN)\left(p + \frac{aN^2}{v^2}\right)$$

For instance, for CO_2, using as units atmospheres, liters, and degrees Kelvin and referring to one mole $N = 6.022 \cdot 10^{23}$; $Nk = R = 0.08206$; $N^2 a = 3.592$; $Nb = 0.04267$.

For the zone in the pv plane in which liquid and gas coexist, van der Waals isothermals are replaced by horizontal straight lines corresponding to the pressure of the saturated vapor. These lines intercept the van der Waals isothermals at three points and form two loops with the same. Where are these horizontal lines to be located? Maxwell gave the criterion that the two loops must have the same area. His thermodynamical argument considers a cycle formed by the horizontal line and the van der Waals isothermal. If we follow this cycle, we go around the two loops, leaving one to the right and one to the left. The loops thus have opposite signs. Consider now the cycle mentioned above. It is completely isothermal; thus, it has efficiency zero, and no mechanical work can be performed. The mechanical work is given by the area of the cycle, which thus must be zero. Hence the two loops must have equal areas.

Bibliography

(Excerpts found in the text are taken from books in the bibliography.)

The general bibliography reported here is limited to the most important works. In many of the books cited, there are extended lists of references, as well as references to original papers.

For dictionaries and encyclopedic works, see C. C. Gillispie, ed., *Dictionary of Scientific Biography,* 16 vols. (New York: Scribners, 1970–1978).

H. Buckley, *A Short History of Physics* (London: Methuen, 1929) is exactly what the title says, and I have read it with pleasure. The series by E. Mach — *The Principles of Mechanics* (repr. LaSalle, Ill.: Open Court, 1960), *The Principles of Physical Optics* (repr. New York: Dover, 1953), and *Die Principien der Wärmelehre, historisch-kritisch entwickelt* (Leipzig: Barth, 1896) — contains much original material presented in the author's personal perspective; the books were written about a century ago.

H. A. Boorse and L. Motz, eds., *The World of the Atom* (New York: Basic Books, 1966) is a good anthology with ample biographical notes; it generally refers to the twentieth century, but it has valuable references to earlier times. E. Whittaker, *A History of the Theories of Aether and Electricity* (New York: Harper & Row, 1960), is an erudite work with excellent mathematical developments. F. W. Ostwald, *Klassiker der exacten Naturwissenschaften,* is a collection of reprints of fundamental papers.

French science at the time of Napoleon and later is vividly portrayed in the biographies contained in F. Arago, *Oeuvres* (Paris: T. Margand, 1865), and M. Crosland, *The Society of Arcueil* (Cambridge, Mass.: Harvard University Press, 1967).

J. G. Crowther, *British Scientists of the Nineteenth Century* (London: Routledge, 1962), is a first-class popular account. P. M. Harman, *Energy, Force, and Matter: The Conceptual Development of Nineteenth-century Physics* (Cambridge: Cambridge University Press, 1982), is a very compressed summary of the subject, with an excellent bibliographical essay.

Chapter 1: The Founding Fathers: Galileo and Huygens

The literature on Galileo is immense. Here I give only a few books; the works cited also contain much bibliographical information.

Fundamental is A. Favaro, ed., *Le opere di Galileo Galilei,* in the National Edition, 20 vols. (Florence: Barbera, 1890–1909), with additions. Favaro's many books and articles are all of high quality.

Valuable English translations are *Dialogue Concerning the Two Chief World Systems, Ptolemaic and Copernican* by S. Drake (Berkeley: University of California Press, 1967) and an anthology of Galileo's works in English translation is S. Drake, *Discoveries and Opinions of Galileo* (New York: Doubleday, 1957).

Biographies include L. Geymonat, *Galileo Galilei,* translated by S. Drake (New York: McGraw-Hill, 1965), and S. Drake, *Galileo at Work, His Scientific Biography* (Chicago: University of Chicago Press, 1978). A modern Catholic work is P. Paschini, *Vita e opere di Galileo Galilei* (Rome: Pontifical Academy of Sciences, 1965).

The collected works of Huygens, *Oeuvres complètes de Christiaan Huygens,* 22 vols. (The Hague: M. Nijhoff, 1888–1950), is the fundamental source of information.

See also *Studies on Christiaan Huygens (A Symposium)* (Lisse: Swets & Zeitlinger, 1980); A. E. Bell, *Christiaan Huygens and the Development of Science in the Seventeenth Century* (London: Longmans, Gren, 1947); and S. P. Thompson's translation of *Treatise on Light* (repr. New York: Dover 1952).

Chapter 2: The Magic Mountain: Newton

A modern edition of the *Principia* in English is *Sir I. Newton's Mathematical Principles of Natural Philosophy and His System of the World,* translated by A. Motte (1729) and edited by F. Cajori (Berkeley: University of California Press, 1934). See also Newton's *Opticks* (repr. New York: Dover, 1952).

The classic biography is D. Brewster, *Memoirs of the Life, Writings, and Discoveries of Isaac Newton,* 2 vols. (Edinburgh: Constable, 1855). A modern biography is L. T. More, *Isaac Newton, A Biography* (New York: Scribners, 1934). The most recent scholarly biography, with much new information and bibliographical addenda, is R. S. Westfall, *Never at Rest* (Cambridge: Cambridge University Press, 1980). F. E. Manuel, *A Portrait of I. Newton* (Washington: New Republic Books, 1980), is a psychoanalytical essay on Newton.

The article by I. B. Cohen, "Newton in the Light of Recent Scholarship," *Isis,* 51 (1960), 489–514, gives an idea of recent work on Newton. See also D. T. Whiteside, "The Expanding World of Newtonian Research," *History of Science,* 1 (1962), 16–29.

Chapter 3: What Is Light?

On Thomas Young, see A. Wood, *Thomas Young, Natural Philosopher, 1773–1829* (Cambridge: Cambridge University Press, 1954), and G. Peacock and

J. Leitch, eds., *Miscellaneous Works of the Late Thomas Young,* 3 vols. (London: Murray, 1855).

For Fresnel, see H. de Sénarmont, S. D. Poisson, É. Verdet, L. Fresnel, eds., *Oeuvres complètes d'Augustin Fresnel,* 3 vols. (Paris: Imprimerie Imperiale, 1866–1870), which is of fundamental importance. It also contains letters and biographical material.

See also M. Métivier, P. Costabel, P. Dugac, eds., *S. D. Poisson et la science de son temps* (Palaiseau: École Polytechnique, 1981).

For Fraunhofer, see M. v. Rohr, *J. Fraunhofer's Leben, Leistungen und Wirksamkeit* (Leipzig: Akademisches Verlag, 1929), and E. C. J. Lommel, ed., *Joseph von Fraunhofer gesammelte Schriften* (Munich: Verlag der K. Akademie, 1888).

Chapter 4: Electricity: From Thunder to Motors

For early studies of electricity and an excellent essay on eighteenth-century physics, see J. L. Heilbron, *Electricity in the Seventeenth and Eighteenth Centuries* (Berkeley: University of California Press, 1979).

For Volta there is the National Edition of *Le opere di Alessandro Volta,* 7 vols. (Milan: 1918–1929). A good biography is G. Polvani, *Alessandro Volta* (Pisa: Domus Galilaeana, 1942).

For Franklin, the *Autobiography* is not too informative on his scientific work, but is most interesting in other respects. See also C. Van Doren, *Benjamin Franklin* (New York: Greenwood, 1938), and I. B. Cohen, *Benjamin Franklin, Scientist and Statesman* (New York: Scribners, 1975).

See also S. C. Brown, *Benjamin Thompson, Count Rumford* (Cambridge, Mass.: MIT Press, 1981), and H. Hartley, *Humphry Davy* (London: 1967).

For Faraday, see his *Experimental Researches in Electricity,* 3 vols. (London: Taylor, 1839–1855). See also T. Martin, ed., *Faraday's Diary,* 7 vols. (London: Bell, 1932–1936). M. Faraday, *The Chemical History of a Candle* (Chicago: Chicago Review Press, 1980), is a reprint of a classic, showing Faraday's lecturing style. Biographies include H. Bence Jones, *Life and Letters of Faraday,* 2 vols. (London: Longmans, Green, 1870), [the author was the secretary of the Royal Institution at the time of Faraday] and L. P. Williams, *Michael Faraday, A Biography* (London–New York: 1965).

For Maxwell, see W. D. Niven, ed., *The Scientific Papers of James Clerk Maxwell,* 2 vols. (repr. New York: Dover, 1952); J. C. Maxwell, *A Treatise on Electricity and Magnetism,* 2 vols. (Oxford: Clarendon Press, 1891); J. C. Maxwell, *Theory of Heat* (London: Longmans, Green, 1870); and L. Campbell and W. Garnett, *The Life of James Clerk Maxwell* (London: Macmillan, 1882). C. W. F. Everitt, *James Clark Maxwell, Physicist and Natural Philosopher* (New York: Scribners, 1975) is concise and excellent and contains an extended bibliography. See also L. Rosenfeld, "The Velocity of Light and the Evolution of Electrodynamics," *Suppl. Nuovo Cimento,* 4 (1956), 1630.

J. Hertz, *H. Hertz* (Leipzig: Akademische Verlagsgesellschaft, 1927), is a delightful, vivid, and first-hand account.

An amusing fantasy worth reading is R. McCormmach, *Night Thoughts of a Classical Physicist* (Cambridge, Mass.: Harvard University Press, 1982).

On Althoff, see A. Sachse, *Friedrich Althoff und sein Werk* (Berlin: E. S. Müller, 1928).

For Lorentz, see G. L. de Haas-Lorentz, ed., *H. A. Lorentz: Impressions of His Life and Work* (Amsterdam: North-Holland, 1957).

Chapter 5: Heat: Substance, Vibration, and Motion

For Sadi Carnot, see E. Mendoza's edition of *Reflections on the Motive Power of Fire* (repr. New York: Dover 1960), which also contains fundamental papers by Clapeyron and Clausius. *Sadi Carnot et l'éssor de la thermodynamique* (Paris: École Polytechnique, 1967) is a symposium on various aspects of Carnot's life and work.

For Lord Kelvin's work, see *Mathematical and Physical Papers,* 6 vols. (Cambridge: Cambridge University Press, 1882–1911). Biographies include S. P. Thompson, *The Life of Lord Kelvin* (London: Macmillan, 1901); A. Gray, *Lord Kelvin* (London: J. M. Dent, 1908); and H. I. Sharlin, *Lord Kelvin: The Dynamic Victorian* (University Park: Pennsylvania State University Press, 1979). See also J. Larmor, "Obituary Notice of Lord Kelvin," *Proceedings of the Royal Society,* **81** (1908). G. Green and J. T. Lloyd, *Kelvin's Instruments and the Kelvin Museum* (Glasgow: University of Glasgow, 1970) presents in a graphic way the instruments built under Kelvin's direction, illustrating strengths and weaknesses of the experimental art of his times.

For Mayer's works, see J. J. Weyrauch, ed., *Die Mechanik der Wärme,* 3rd ed., (Stuttgart: Cotta, 1893). S. Friedländer, *Julius Robert Mayer* (Leipzig: T. Thomas, 1905), is a good biography.

For Helmholtz's papers, see *Wissenschaftliche Abhandlungen von Hermann Helmholtz,* 3 vols. (Leipzig: Barth, 1882–1894), and an anthology in English, R. Kahl, ed., *Selected Writings of Hermann von Helmholtz* (Middletown, Conn.: Wesleyan University Press, 1971). For a biography, see L. Koenigsberger, *Hermann von Helmholtz,* translated by F. A. Welby (New York: Dover, 1956).

For Clausius, see F. Folie, "R. Clausius. Sa vie, ses travaux et leur portée métaphysique," *Revue des questions scientifiques* (Bruxelles), **27** (1890), 419; and M. J. Klein, "Gibbs on Clausius," *Historical Studies in the Physical Sciences,* 1 (1969), 127.

Illuminating of the period are *Werner von Siemens, Inventor and Entrepreneur: Recollections* (London: Lund Humphries, 1966), and R. L. Stevenson, "Memoir of Fleeming Jenkin," in *Papers of Fleeming Jenkin* (London: Longmans, Green, 1887).

Chapter 6: Kinetic Theory: The Beginning of the Unraveling of the Structure of Matter

An excellent, modern, and scholarly history of the subject is S. G. Brush, *The Kind of Motion We Call Heat,* 2 vols. (Amsterdam: North-Holland, 1976). See also S. G. Brush, *Kinetic Theory,* 2 vols. (New York: Pergamon Press, 1966), which contains a collection of selected readings with valuable comments.

For Boltzmann's papers, see F. Hasenöhrl, ed., *Wissenschaftliche Abhandlungen* (Leipzig, 1909; repr. New York: Chelsea, 1968). *Populäre Schriften* (Leipzig: Barth, 1905), contains lectures and articles. See also E. Broda, *Ludwig Boltzmann: Mensch, Physiker, Philosoph* (Berlin: F. Deuticke, 1955), and *International Symposium, 100 Years of Boltzmann's Equation, Vienna, 1972* (New York: Springer, 1973).

For van der Waals, see M. J. Klein, "The Historical Origins of the van der Waals Equation," *Physics,* 73 (1974), 28.

For Gibbs, see H. A. Bumstead and R. G. Van Name, eds., *The Scientific Papers of J. Willard Gibbs,* 2 vols. (New York: Dover, 1961), and L. P. Wheeler, *Josiah Willard Gibbs, The History of a Great Mind* (New Haven: Yale University Press, 1951).

For Maxwell and Thomson, see the bibliographies for Chapters 4 and 5.

Name Index

Subject Index